# NEIGHBORHOOD PLANNING AND COMMUNITY-BASED DEVELOPMENT

4/28/00

# Cities & Planning Series

The *Cities & Planning Series* is designed to provide essential information and skills to students and practitioners involved in planning and public policy. We hope the series will encourage dialogue among professionals and academics on key urban planning and policy issues. Topics to be explored in the series may include growth management, economic development, housing, budgeting and finance for planners, environmental planning, GIS, small-town planning, community development, and community design.

## Series Editors

Roger W. Caves, Graduate City Planning Program,
   *San Diego State University*

Robert J. Waste, Department of Political Science,
   *California State University at Chico*

Margaret Wilder, Department of Geography and Planning,
   *State University of New York at Albany*

## Advisory Board of Editors

William Peterman

# NEIGHBORHOOD PLANNING AND COMMUNITY-BASED DEVELOPMENT

The Potential and Limits
of Grassroots Action

Cities & Planning

**SAGE** Publications
*International Educational and Professional Publisher*
Thousand Oaks   London   New Delhi

*For information:*

Sage Publications, Inc.
2455 Teller Road
Thousand Oaks, California 91320
E-mail: order@sagepub.com

Sage Publications Ltd.
6 Bonhill Street
London EC2A 4PU
United Kingdom

Sage Publications India Pvt. Ltd.
M-32 Market
Greater Kailash I
New Delhi 110 048 India

Printed in the United States of America

*Library of Congress Cataloging-in-Publication Data*

Peterman, William.
    Neighborhood planning and community-based development:
The potential and limits of grassroots action / by William Peterman.
        p. cm.—(Cities and planning)
    Includes bibliographical references and index.
    ISBN 0-7619-1198-7 (cloth: alk. paper)
    ISBN 0-7619-1199-5 (pbk.: alk. paper)
    1. Community development, Urban. 2. Community development,
Urban—Illinois—Chicago—Case studies. 3. Urban renewal. 4. Urban
renewal—Illinois—Chicago. 5. Neighborhood. 6. Neighborhood—Illinois—
Chicago—Case studies. I. Title. II. Cities & planning series.
HN90.C6 P49 2000
307.3'36216--dc21                                                    99-050429

This book is printed on acid-free paper.

00   01   02   03   04   05   06   7   6   5   4   3   2   1

| | |
|---|---|
| *Acquisition Editor:* | Harry Briggs |
| *Editorial Assistant:* | Mary Ann Vail |
| *Production Editor:* | Sanford Robinson |
| *Editorial Assistant:* | Cindy Bear |
| *Typesetter:* | Christina M. Hill |
| *Indexer:* | Teri Greenberg |
| *Cover Designer:* | Ravi Balasuriya |

# CONTENTS

# PREFACE

I first became involved with neighborhood planning and community development when I was on the faculty of Bowling Green State University in Ohio during the 1970s. Almost by chance I developed an association with the Toledo Metropolitan Mission (TMM). TMM had received a small grant to assist an ad hoc group of citizens on the fringe of the urbanized area of Toledo in an effort to fight urban sprawl.[1] I first inquired about the effort and then volunteered to help. At the time, TMM, the action arm of the Toledo (Ohio) Council of Churches, was headed by an ex-priest, Jerry Ceille, who had previously been an associate of the well-known Milwaukee activist priest, Father Groppi. It was through working with TMM and Ceille that I initially came to discover how community organizing, activism, and community-oriented research could bring about effective community building.

Building on my initial work with TMM, I applied for and received a National Science Foundation "public service science residency." The short-lived residency program (I believe it lasted only 2 years) allowed me to take a year's leave of absence to work full-time with TMM. During this year, I worked on several neighborhood issues being supported by TMM. I helped a group of churches plan and implement a dial-a-ride medical transportation system for inner-city seniors, helped another group of seniors develop plans for an activity center, and helped found the Greater Toledo Housing Coalition. The highlight of this last activity was traveling to Cincinnati to present the case for the very first challenge ever made under the then newly enacted Community Reinvestment Act (CRA).

After my residency was over, I continued to work with TMM on a voluntary basis, but in 1979 I got the opportunity to go to the University of Illinois at Chicago and to become the director of the newly formed Voorhees Center for Neighborhood and Community Improvement, a unit of the College of Architecture, Art, and Urban Planning.[2] The search committee for the director was looking for an academic with experience in doing community development, and my work with TMM had made me somewhat unique among those who had applied. In a written document in which I articulated what I hoped to be able to do as director of the Voorhees Center, I argued that research and technical assistance were important for supporting community organizing and community development. I also stated that I believed that universities had an obligation to serve the needs of the communities in which they are located and that an outreach center was one way that this obligation could be met.

During the approximately 11 years that I directed the Voorhees Center, I had the opportunity to work in a variety of Chicago's communities, with many different community organizations, and on a variety of topics ranging from an inventory of art owned by the Chicago Park District to the opposition of a sports stadium complex on Chicago's near west side. In each instance, the Voorhees Center responded to a request for assistance from the neighborhood and partnered on projects with community-based organizations or areawide organizations concerned about neighborhood development. A few of these efforts are documented in this book.

I have always considered what we did at the Voorhees Center to be research in that we were constantly seeking to find new ways and means to improve the quality of neighborhoods and the socioeconomic well-being of neighborhood people.[3] We, ourselves, did not become directly involved in community organizing but instead worked alongside of individuals whose mission was community organizing. It was the dual efforts of research and organizing that resulted in most of our successes. It should be noted, however, that neighborhood work of this type is not easy, and our failures were as numerous as our successes.

In 1991, I left the Voorhees Center to become a regular faculty member in the University of Illinois at Chicago's School of Urban Planning and Policy.[4] Then in 1995, the opportunity arose for me to apply to head up yet another urban center, this time at Chicago State University, a medium-sized comprehensive teaching university on Chicago's south side. I applied and was offered the position. So now, once again, I am attempting to bring the resources of a university to bear on the problems of urban neighborhoods.

I began writing this book with encouragement from several of my colleagues, and I hope that I can live up to their expectations. In many ways, I no longer believe as I did 15 to 20 years ago that community-based efforts are the only way to do urban development. Nor do I believe that community-based organizations are always fair, equitable, and right. And although I am hopeful about the future of our cities, I am not optimistic about current trends relating to poverty, immigration, affirmative action, and overall fairness.

I believe that to do successful neighborhood work, one must be a realist. Global, national, and local forces have significant impacts on what happens in neighborhoods, and no amount of community organizing at the local level will change these forces. Racism and sexism cannot be countered by working only at the neighborhood level. But unorganized and unassisted communities, whether they be communities of place or communities of common interest or identity, will always suffer more than those that have come together to determine what they want and expect and have identified and obtained the resources needed to at least make progress toward their visions. On my office wall hangs a small poster with a quote from the legendary labor activist Mother Jones: "Pray for the dead and fight like hell for the living" (Jones 1925, p. 41). I have always tried to follow Mother's charge.

It is argued that cities have always been the center of civilization (Mumford, 1961). Despite some progress toward revitalizing some parts of some cities, our metropolitan areas remain in trouble, fragmented politically and socially, with too many poor and disadvantaged people compelled to live in deteriorating areas far from jobs and other opportunities. We tend to call places where the poor and disadvantaged live neighborhoods, although many of these places have few if any neighborhood amenities. But it seems to me that as our neighborhoods die, so do the cities in which they exist, and ultimately the entire metropolis suffers. So, even though there are significant forces that work against the empowerment of local communities, it is still important and essential to work to improve the quality of life in our central cities. Doing good in central-city neighborhoods will ultimately benefit us all.

## ACKNOWLEDGMENTS

The material in this book covers a period of nearly 20 years. Over those years I worked with numerous people, many who helped me either in my work or in my thinking about neighborhood development. All these people, at least their thoughts and deeds, appear in this book. Although I cannot possibly acknowledge them all, I must at least single out several individuals.

This book would never have been written had it not been for the encouragement of my good colleague, Myron Levine of Albion College. Myron insisted that I put aside my vow to never write a book, and he got Roger Caves, Bob Waste, and Margaret Wilder interested in what I might have to write. They, in turn, got Catherine Rossbach of Sage to show up at my office door one day at Chicago State University. Catherine not only convinced me I should write a book, but she also kept me going throughout the entire process. My wife, Jean Peterman, can also take credit for helping me get to the end. Because her book was published before mine, she showed me, by example, that I could do it.

I have always thought of students as being my colleagues. In the projects that make up the case studies in this book, I was lucky to have had the assistance of several excellent students. Among them were David Browne, Sherrie Hannon, Ken Brierre, Mary Ann Young, and Jean Gunner of the University of Illinois at Chicago and, more recently, Elaine Davis and Allegra Henderson of Chicago State University. Most of my good ideas have resulted from lively discussions with these and other students. I particularly value and miss my conversations with Sherrie Hannon, who was without a doubt my finest student and who died much too soon.

Three individuals have played key roles at various parts of my professional career, and they must be singled out. They are true partners—Hallie Aimee, Stanley Horn, and Shiela Radford-Hill. Their role in shaping my ideas about neighborhood development cannot be underestimated.

Finally, I must acknowledge my current colleagues at Chicago State University—Mark Bouman, Celeste Henderson, and Mike Siola. These three have made it possible for me to continue applying what I know and to help meet at least some of the needs of the communities that make up the southeast and far south sides of Chicago. They also help me to keep my spirits up when things are not going just right. That I could write this book while expanding the outreach work at Chicago State University suggests that they know their jobs and do them well, sometimes despite my interference.

## NOTES

1. My wife was on the committee that awarded the grant to TMM. Following one of the committee meetings, she mentioned the project to me as one that I might be interested in. I subsequently contacted TMM about it. Although skeptical at first about working with some unknown young academic, TMM's director, Jerry Ceille, eventually accepted my offer to assist in the project on a pro bono basis.

2. Along with the Center for Urban Economic Development (CUED), the Voorhees Center became the model on which the university's "Great Cities" program would be developed in the 1990s. At the time I became director of the Voorhees Center, there was little support for grassroots activism on the part of the University of Illinois at Chicago administration, and the center initially got a poor review because it did too little "scholarly" work and did not bring enough grant money into the university.

3. Sometimes what we did would result in work that could be published in academic journals, but often it did not. Publishing and grant making were not the primary goals of the center, and this occasionally got us in trouble with our academic colleagues.

4. I was followed as director of the Voorhees Center by Patricia Wright, who had been on the staff of the CUED at the University of Illinois at Chicago. Under Pat's guidance, the Voorhees Center has continued to assist community-based efforts throughout Chicago and, in my personal opinion, has carried the work of the center to a new and higher level.

# CHAPTER 1

# INTRODUCTION

This book is about neighborhoods, neighborhood-based grassroots development efforts, and the role that planning can and does play in strengthening neighborhood development efforts. It accepts the notion that grassroots efforts make a difference, it explores the conditions under which they succeed, but it also identifies and discusses the limitations that can arise from focusing too narrowly on the neighborhood as the unit of analysis and context for the solution of urban problems.

Neighborhoods and grassroots community-based efforts have not always been viewed as central to the redevelopment or revitalization of our cities. In the late 1960s and early 1970s, when I was a graduate student studying the city and its problems, my professors talked a lot about the federal government and its role in providing leadership, programs, and resources to address problems of urban development. At that time, urban policy, whether it dealt with regional growth or neighborhood development, emanated for the most part at the federal level. This was the way it had been since the Great Depression of the 1930s, when the federal government had first paid serious attention to the plight of cities.

Responding to a need to get the construction industry back on its feet, the Roosevelt administration fashioned an urban policy beginning in 1934 that supported financial institutions through the creation of the Federal Housing Administration (FHA) and then, beginning in 1937, provided direct subsidies for housing production through the creation of the public housing program (see Bratt, 1989, pp. 18-24). Following World War II, these programs were expanded. FHA lending policies fueled suburbanization, and central-city neighborhoods were red-lined. "Urban renewal" devastated large urban areas and necessitated the construction of high-rise public housing to accommodate those households that were displaced. Eventually, urban renewal evolved into the programs of the "Great Society" of the 1960s, culminating with the Demonstration Cities and Metropolitan Development Act of 1966 (Model Cities). At this point, attention began to shift away from feder-ally mandated approaches and toward programs fashioned at the neighborhood level.

Federal urban policy following World War II had three general ob-jectives: the elimination of slums, the reuse of land for middle-class housing and institutional expansion, and the revitalization of city cores. The results, however, were the destruction of neighborhoods, the ghettoization of the poor and minorities into public housing, and the continued decline of central cities (see Hirsch, 1983).

Beginning with President Nixon's moratorium on federal housing subsidies in 1972, the federal role in addressing urban problems began to decline. The passage of the Housing and Community Development Act of 1974 shifted the focus of urban problem solving away from the federal government and toward local municipalities. In turn, munici-palities by and large passed the responsibility along to organizations operating in the urban neighborhoods where the problems existed.

New kinds of community-based organizations and community development corporations (CDCs) sprung up to use the dollars that were flowing to the neighborhoods. Although for the most part com-munity initiated and controlled, CDCs have always had a connection both to government funding and to private sources of capital. Their roots can be traced to funding provided by a Special Impact Amend-ment to the Economic Opportunity Act (Bratt, 1989). The first CDC, the

Bed-Stuy Restoration Corporation, was founded following a visit by Robert Kennedy to the devastated Bedford-Stuyvesant section of New York's Central Brooklyn.

CDCs differed from advocacy-oriented organizations in that they were designed to operate more like a business and were expected to carry out revitalization projects by working with, rather than in opposition to, local government and private institutions such as banks. Although the initial thrust of CDCs was in the area of job creation, by the 1970s most CDCs had turned to the task of building new housing or renovating existing housing and renting it to low- and moderate-income households (Stoecker, 1997). Some CDCs also attempted a variety of economic initiatives such as opening a grocery store or a small factory that employed local residents.

Shifting the focus of development to the local level meant that the community, rather than the federal government, came to be seen as the unit of solution for urban problems (Checkoway, 1984). Nearly everyone involved with urban policy, government officials, neighborhood activists, journalists, planners, and academics applauded this shift. They argued that locally directed development placed the emphasis on solving urban problems where it rightly belonged—on the community. Locally focused efforts, it was believed, promoted self-determination and empowerment of neighborhood residents, who too often in the past had been the victims of urban policies.

But shifting the focus to the community was not the only change in urban policy at the federal level. As the federal government began to take less direct responsibility for central-city neighborhood programs, it also began to provide fewer of the resources those neighborhoods desperately needed, particularly the dollars needed to plan and implement meaningful redevelopment strategies. This left many neighborhoods with control over revitalization programs but too few resources to accomplish what was needed.

Despite a near consensus for at least 20 years that communities or neighborhoods are the appropriate unit out of which to fashion a solution to urban problems and that an empowered community can bring about its own renaissance, there are far too few success stories of grassroots self-determination leading to neighborhood revitalization. The number of neighborhoods in our cities that are failing still outnumbers those that are succeeding, and when a neighborhood does revitalize, it most often does so by gentrifying, which displaces many of the existing residents and merely shifts the neighborhood's problem to some other neighborhood.

The somewhat limited success of grassroots neighborhood development should give us pause to reflect on whether it is an idea that can live up to the challenges given current social, political, and environmental realities. Because neighborhood development assumes the existence of places called neighborhoods, we should perhaps first revisit existing theories of neighborhoods to see if they are still relevant in the context of urban development at the end of the 20th century. Then we can turn to theories of neighborhood development to see if they address the issues and needs of the modern "neighborhood."

## ORGANIZATION OF THE BOOK

In this book, I explore either directly or indirectly some commonly held assumptions about neighborhoods and neighborhood development. I question whether the concept of neighborhood, as defined by urban sociologists and urban planners, has meaning in large cities. I also question to what extent traditional notions about neighborhoods guide what planners and neighborhood activists do in their attempts to bring about neighborhood redevelopment.

I also critically explore the possibilities and limits of bottom-up grassroots strategies of community organizing, development, and planning as a means for successfully maintaining and revitalizing urban neighborhoods. In doing so, I also identify and describe the conditions that I believe are necessary to the success of grassroots efforts.

I support grassroots neighborhood-based efforts to revitalize urban communities and have worked most of my adult life to assist such efforts. I believe that the community is or should be the unit of solution.

Yet, I also believe that communities, acting by themselves, cannot really solve their problems. Although significant amounts of both economic and human capital exist in many so-called "blighted neighborhoods," I contend that this capital alone is insufficient to overcome the systemic forces that work to impoverish some neighborhoods and to enrich others. Community empowerment may be at the heart of neighborhood development, but it is not the only thing that is necessary.

To accomplish my purposes, I draw on my more than 20 years of experience in doing planning with an advocacy focus—first in Toledo, Ohio, then as the director of a neighborhood outreach center at the University of Illinois at Chicago, and most recently at Chicago State University. The heart of the book, Chapters 5 through 8, is an account of four community-based efforts to control or stimulate neighborhood development. Each of these is an effort in which I have had some involvement. They are also efforts in which the degree of success (or failure) varies. Taken as a whole, I believe, they help to clarify both the potential and limits of community development that can come from grassroots organizing, community-based development, and neighborhood planning. They also help clarify the conditions necessary for successful development.

The case studies, however, must be placed in the context of the meanings of commonly used terms, such as *neighborhood* and *community* as well as *community empowerment.* These are terms often used by activists and academics alike, but they are rarely defined. They lie, however, at the base of our efforts to do neighborhood planning and community development, and thus our understanding of what we mean by them is critical to our efforts as planners to bring about meaningful urban revitalization.

The next chapter thus begins with a discussion of the meanings of neighborhood and community. The questions addressed include the following:

What do we mean by the use of the terms *neighborhood* and *community?*
What are the major perspectives on the goals and objectives of community organizing and community-based development initiatives?
What are the alternative approaches to doing neighborhood planning?

Chapter 3 begins with a discussion of the notion of empowerment. In the chapter, I argue that empowerment means different things, depending on the ideological context in which the term is used, and that actions consistent with conservative or liberal meanings do not lead to a kind of community development that is fully sensitive to resident hopes and desires. An alternative meaning, one that I associate with a progressive ideology, will be offered as one that is more supportive of true community development.

This discussion about empowerment is followed by a review of the basic notions behind citizen participation and community organizing and community organizations. The chapter concludes with a discussion about CDCs and their role in bringing about neighborhood change and community empowerment.

Chapter 4 focuses on the changing dynamics and structure of urban places, with special emphasis given to changes that have occurred in Chicago over the past several decades. It is against the backdrop of urban growth and decline that neighborhood redevelopment efforts must occur and, as such, it becomes the backdrop for the case studies in this book.

The next four chapters (Chapters 5-8) present the case studies of specific efforts to bring about some type of neighborhood revitalization in Chicago. Figure 1.1 is a map of Chicago, showing the location of each neighborhood. The neighborhoods are scattered throughout the city—one on the near south side, one on the far south side, one southwest, and one on the north side. Public housing is present in two of the neighborhoods, and it plays a prominent role in the revitalization effort. The other two neighborhoods contain mixtures of rental and owner housing consisting of single-unit dwellings and small apartment buildings. All are representative of typical neighborhoods found in Chicago.

In each of the case studies, a neighborhood group, responding to a real or implied challenge to neighborhood stability, attempts to organize and plan for neighborhood betterment. In Chapter 4, the issue is gentrification. In Chapter 5, the issue is the threat of physical destruction. In Chapters 6 and 7, the issue is community viability, stemming deterioration in the first case and dealing with an already deteriorated community in the second. The neighborhoods discussed in each case continue to exist today, although not fully in the condition envisioned by the community and those of us who attempted to help it achieve its desired future.

Four conditions, which I conclude are necessary to fully empower a community in ways that result in real grassroots-based development, are then presented in Chapter 9. These conditions are a refinement of those initially articulated by Daniel Monti (1989) in his study of

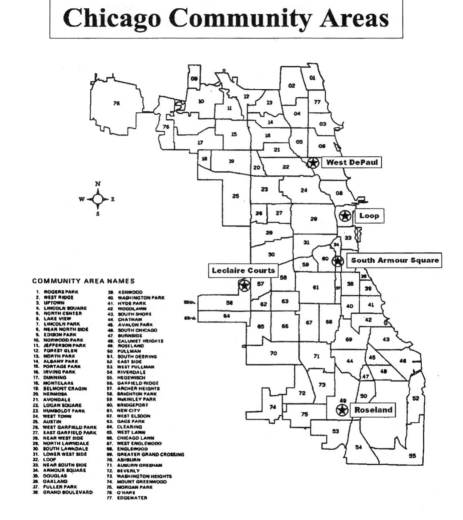

# Chicago Community Areas

**COMMUNITY AREA NAMES**

1. ROGERS PARK
2. WEST RIDGE
3. UPTOWN
4. LINCOLN SQUARE
5. NORTH CENTER
6. LAKE VIEW
7. LINCOLN PARK
8. NEAR NORTH SIDE
9. EDISON PARK
10. NORWOOD PARK
11. JEFFERSON PARK
12. FOREST GLEN
13. NORTH PARK
14. ALBANY PARK
15. PORTAGE PARK
16. IRVING PARK
17. DUNNING
18. MONTCLARE
19. BELMONT CRAGIN
20. HERMOSA
21. AVONDALE
22. LOGAN SQUARE
23. HUMBOLDT PARK
24. WEST TOWN
25. AUSTIN
26. WEST GARFIELD PARK
27. EAST GARFIELD PARK
28. NEAR WEST SIDE
29. NORTH LAWNDALE
30. SOUTH LAWNDALE
31. LOWER WEST SIDE
32. LOOP
33. NEAR SOUTH SIDE
34. ARMOUR SQUARE
35. DOUGLAS
36. OAKLAND
37. FULLER PARK
38. GRAND BOULEVARD
39. KENWOOD
40. WASHINGTON PARK
41. HYDE PARK
42. WOODLAWN
43. SOUTH SHORE
44. CHATHAM
45. AVALON PARK
46. SOUTH CHICAGO
47. BURNSIDE
48. CALUMET HEIGHTS
49. ROSELAND
50. PULLMAN
51. SOUTH DEERING
52. EAST SIDE
53. WEST PULLMAN
54. RIVERDALE
55. HEGEWISCH
56. GARFIELD RIDGE
57. ARCHER HEIGHTS
58. BRIGHTON PARK
59. McKINLEY PARK
60. BRIDGEPORT
61. NEW CITY
62. WEST ELSDON
63. GAGE PARK
64. CLEARING
65. WEST LAWN
66. CHICAGO LAWN
67. WEST ENGLEWOOD
68. ENGLEWOOD
69. GREATER GRAND CROSSING
70. ASHBURN
71. AUBURN GRESHAM
72. BEVERLY
73. WASHINGTON HEIGHTS
74. MOUNT GREENWOOD
75. MORGAN PARK
76. O'HARE
77. EDGEWATER

**Figure 1.1.** Map of Chicago Showing the Locations of the Neighborhoods in the Four Case Studies

resident-managed public housing. I contend that even though Monti developed the conditions as appropriate only in the narrow instance of a public housing community, they are, with some modification, valid over a broad range of community development and neighborhood planning situations.

After presenting and discussing the conditions necessary for community empowerment, I return briefly to each of the four case studies to show how the conditions were or were not met in each case. My argument is that the successes or failures in each case—that is, the successes or failures of reaching the desired community goals and objectives—can be accounted for by attention or lack of attention to the four conditions.

The concluding chapter of the book is an attempt to assess and summarize what planners and other urban specialists can expect to accomplish through a community-based approach to planning. Ideas as to how planning proceeds in a neighborhood setting and the conditions under which it can succeed are presented as a "tool kit," intended to provide planners and community leaders with appropriate and useful procedures for doing neighborhood-based planning and development.

This book is intended to be of interest to a variety of readers. It should be of special interest to individuals who are directly involved in neighborhood planning and development activities. This includes not only public planners and planners with not-for-profit community-oriented agencies and organizations but also staff of community development corporations and social service agencies concerned about neighborhood and community viability. Community organizers should also find the book useful in their efforts to create viable and active neighborhood organizations. Planning students and educators should also be interested in the book because it not only shows how neighborhood planning is done, but it also lays out at least a rudimentary theory of the neighborhood development process and raises basic questions about notions of neighborhood and community. And finally, individuals simply interested in neighborhoods and neighborhood organizations and how they work may find the case studies in Chapters 5 through 8 interesting reading because they tell the stories of real people in real situations and depict the struggles that people undertake in attempting to preserve and revitalize their neighborhoods and communities.

# 2

# NEIGHBORHOODS, COMMUNITIES, COMMUNITY DEVELOPMENT, AND NEIGHBORHOOD PLANNING

In the first part of this chapter, I explore the notions of neighborhood and community as they are commonly used by planners, community developers, and neighborhood people themselves. This is followed by a similar discussion of the notions of neighborhood planning and how it is carried out. The chapter serves to lay out the context for a discussion of efforts to revitalize neighborhoods through community development and thus sets the stage for the case studies presented in Chapters 5 through 8.

## THE VARIED MEANINGS OF
## NEIGHBORHOOD AND COMMUNITY

The terms *neighborhood* and *community* are commonly used in conversation, speech, and writing to describe certain urban places as though everybody knows and agrees as to what they mean. Yet, as I will argue, there does not seem to be a consensus as to just what a neighborhood or a community is. Furthermore, the common images associated with these words may be inappropriate as a model for doing meaningful neighborhood planning or community development.

More than simple academic interest should cause us to be concerned about the meanings of neighborhood and community. Policies, programs, and actions intended to improve the physical, social, cultural, and economic well-being of people living in places called neighborhoods or communities are frequently based implicitly on presumed shared definitions of these terms. Whether these definitions are appropriate and whether they are indeed shared may have a significant impact on what policies, programs, and actions are selected and how we view the outcomes of their implementation.

Quite often, writers and speakers referring to neighborhoods or communities do not tell us what they mean by these words. For the most part, *neighborhood* and *community* exist as undefined terms.[1] Thus, it should not be surprising that people often have different and sometimes contradictory notions as to what they mean, and this can result in different and sometimes contradictory proposals for neighborhood improvement.

Before exploring possible paths to grassroots neighborhood redevelopment, it will be useful, perhaps even necessary, to explore some of the more common meanings of *neighborhood* and the ways in which policymakers, planners, and community activists fashion revitalization strategies based on these meanings. The differences between the meanings of neighborhood and community will also be briefly explored. This will allow us in the next chapter to make sense of the notion of *empowerment,* another term whose meaning often remains undefined, and to consider how neighborhoods or communities can be empowered for action and development through community organization.

## NEIGHBORHOOD AS CONCEPT

In tracing the historical notion of the "neighborhood," we can turn to the writings of Lewis Mumford. In his seminal work, *The City in History* (1961), Mumford argues that by 2000 B.C., the physical characteristics of the city had been created and that the physical structures of the ancient Middle Eastern city would have been familiar to a 19th-century observer. Citing Leonard Woolley's excavations of the ancient Mesopotamian city of Ur, Mumford notes that the ancient city consisted of a "series of more or less coherent neighborhoods in which smaller shrines and temples serve for the householder" (p. 74). The temple, Mumford states, was the focus of the neighborhood unit.

According to Mumford (1961), the neighborhood was a concept familiar to Greek city planners and a component of Greek city plans and planning as it developed during the seventh century B.C. The Greek form of planning, known as *Milesian,* was based on the standard gridiron and divided the city into definite "neighborhoods," which were physical entities that took the shape of superblocks. Mumford contends that this appears to be the "first historic example of a deliberately fabricated neighborhood unit" (p. 193).

However, the neighborhoods of Greek planned cities and later of Roman cities appear to be unlike what we refer to as neighborhoods today in that they appear to primarily have been physical structures, designed to provide a certain orderliness to the overall urban fabric. The social and cultural elements that seem to be embodied in the modern notion of neighborhood appear to have been absent.

This, however, changed in medieval Europe, Mumford (1961) claims. In the medieval city, Mumford states that neighborhoods were "a congeries of little cities, each with a certain degree of autonomy and self-sufficiency, each formed so naturally out of common needs and purposes that it only enriched and supplanted the whole" (p. 310).[2]

We need to question Mumford's (1961) assertions about the historical existence and role of neighborhoods as basic elements of cities. It is likely that Mumford is guilty of "retrospective modernism" (see McIlwain, 1936-1937), that is, viewing the past through criteria applicable, if at all, only in the present. His assertions about neighborhoods are contradicted by Sacks (1989), who has extensively studied medieval

Bristol in England.[3] Sacks states that although the concept of community would have been familiar to the residents of medieval Bristol, they would not have understood it as relating to a bounded social system of place but rather to the more partial, impersonal, and transitory relationships found in society at large. Mumford's "congeries of little cities" may reflect 19th- and 20th-century thought rather than the reality of the 15th and 16th centuries.

Current research seems to suggest that the idea of neighborhood, at least as it is commonly and currently articulated by urban planners, is of more recent origin and that its roots are more likely to be found in the suburbs than in the ancient cities of Ur. Indeed, a review of how the concept of the neighborhood as "urban village" came to be so accepted within the planning profession can lead to a conclusion that neighborhoods are neither as universal nor as integral to urban areas as planners often contend.

According to Southworth and Ben-Joseph (1995), social reformers of the late 19th century saw suburbanization as a vital force and viewed it as a mechanism for rehabilitating the city. Suburbanization was seen as an antidote for urban social ills, which for the most part were seen to be the result of overcrowding and poor sanitation. Thus, originally designed for the upper classes, suburbs also became the model for middle-class developments as well and an ideal to be achieved if possible even in poor communities. This led participants at the First National Conference on City Planning and the Problems of Congestion, held in Washington, D.C., in 1909, to call for encouraging developers to build at the edge of cities to relieve congestion and for establishing regulations for proper land subdivision (Southworth & Ben-Joseph, 1995).

Many urban social reformers at the beginning of this century were influenced by the utopian garden city model put forward by Ebenezer Howard (1902/1945). Howard proposed an urban design that he felt incorporated the best of both the city and the country. His goal was to eliminate both the evils of the city and the bucolic boredom of the countryside and to create a self-sufficient village within a city. Howard placed culture and commerce at the center of his urban unit surrounded by residences. The residential area, in turn, was surrounded by a "green belt" through which paths and roads led to factories. Thus, the workplaces were accessible to but removed from the residential areas, which at least partially solved the problems of pollution and other negative externalities associated with industrial activity.

Howard, himself, had been influenced by the efforts of George Pullman to create a "worker's paradise" and had visited what may be one of the best examples of a fully developed garden city in the United States. The town of Pullman, for which construction had begun in 1880 on open land south of Chicago, was designed by the well-known architect Solan Beman, following the railroad car industrialist Pullman's instructions. Pullman Town contained a variety of housing types ranging from the plant manager's stately home (currently used as a restaurant) to dormitories for unmarried workers. All were in close proximity to each other and to the Pullman Palace Car Factory. Everyone could walk to work, partake of cultural offerings, and recreate along the shore of Lake Calumet, which bordered the town (Smith, 1995).[4]

Many urban residential developments at the beginning of this century were a blend of Howard's (1902/1945) garden city concept and a reworking of Olmsted and Vaux's 1868 plan for the Chicago suburb of Riverside (Langdon, 1994). Like Howard's garden city, the plan of Riverside attempted to bring the countryside into the city but did so by creating large lots, winding streets, and a "common" or public space for recreation and location of public buildings such as a village hall and village library. Spaces were also set aside in the plan for churches as well as schools. The "long common" and other open spaces along the banks of the DesPlaines River provided Riverside with both a rural image and places where local residents could gather for social and other community events.

In reality, however, as Kunstler (1993) has pointed out, Riverside was little more than a real estate development, lacking much of what made Howard's (1902/1945) utopian scheme complete: workplaces, commerce, and housing options. The fathers of Riverside's families worked in downtown Chicago, taking the train to Chicago's Loop each morning and returning in the evening. The train also brought to Riverside those items that the "lady of the house" needed to prepare meals and undertake other domestic chores. And Riverside's houses were required to cost at least $3,000, roughly the annual income of a doctor in 1870. The result was the creation of a socially one-dimensional community (Kunstler, 1993).

The attempt by Olmsted and Vaux to make Riverside a village in a city was copied a generation later by pioneering land developers of the early 20th century. These developers created suburban-like, almost always wealthy subdivisions, and they incorporated into them elements that today are frequently considered to be characteristics of all healthy neighborhoods. In a real sense, these upper-income communities became the norm for the ideal urban neighborhood, and this ideal seems to be held as the model, even in places where there are few if any of the physical characteristics of these planned neighborhoods and in places where the residents are far less wealthy.

The pioneering land developers borrowed from both the garden city and the Riverside models. Because homeownership at the time was limited to the wealthy and near wealthy,[5] these planned developments consisted primarily of single-family detached owner-occupied homes. Land, of course, was set aside for schools and churches. Public parks and playgrounds, however, were often limited or completely absent out of fear that outsiders might be attracted to them. Open spaces were thus usually restricted to small plots of ground for growing flowers or for locating gates or other kinds of identifiable entryways into the community.[6]

The better developments also provided for retail activity. Kansas City developer J. C. Nichols pioneered the concept of the outlying shopping center with his Colonial Shops in 1907, followed in 1923 by the first fully automobile-oriented shopping district in the United States, the Country Club Plaza (Worley, 1990). Like Riverside, Nichols's planned real estate developments omitted the workplace and offered no opportunities for lower-class households.

The developers also were pioneers in creating community organizations, incorporating the concept of the neighborhood association into their new subdivisions. Rudimentary homeowners associations had existed in England since as early as the mid-18th century and in one or two places in the United States by the mid-19th century. But it was J. C. Nichols, beginning in 1910 in Kansas City, who refined the idea and created the first "true" homeowners association (Worley, 1990). Nichols's intent was to get the homeowners to self-police his subdivisions, which he accomplished by transferring the enforcement of deed and other restrictions and the approval of building plans to the association. Homeowner associations remain a feature of suburban subdivisions today and are the model for "civic associations" found in many city neighborhoods. Where they exist in cities, they almost always favor homeownership over renting and gentrification over the status quo.

With the help of Nichols's land development company, the home-owners associations sponsored a variety of social and cultural activities, including lawn decorating competitions, women's clubs, and a model sailboat competition. Although on the surface all these activities appeared to be a means for creating community, their real purpose was to make sure that the homeowners would become a type of neighborhood "police" and that community standards would be maintained (Worley, 1990).

When the Regional Planning Association of America (RPAA) was formed in 1923, its founders, including Lewis Mumford, adopted the garden city ideal of Ebenezer Howard as the model for creating and maintaining the sense of community in residential neighborhoods (Southworth & Ben-Joseph, 1995). Their actual model of the ideal neighborhood, however, more closely resembled the middle- to upper-income real estate developments, such as those of J. C. Nichols, with which they were familiar.

Shortly after the RPAA was formed, Clarence Perry, another founding member, began promoting the concept of "the neighborhood unit" (Perry, 1929) as a "fractional urban unit that would be self-sufficient yet related to the whole" (Southworth & Ben-Joseph, 1995, p. 71). Perry's notion was closely tied to the creation of traffic systems that protected neighborhoods by establishing a hierarchy of streets and keeping main thoroughfares out of existing or planned neighborhoods.

Southworth and Ben-Joseph (1995) credit the introduction of the principles for residential street systems in the New York region, which were laid out by Perry and Thomas Adams in Volume 7 of the New York regional plan of 1929, as being an important step toward the acceptance within the planning profession of the residential neighborhood as a unique entity to be protected and deliberately planned. Many of the physical design characteristics identified by Perry and Adams, including cul-de-sacs, residential streets, and the separation of vehicles and pedestrians, were also incorporated into the 1928 plan of Radburn, designed by Clarence Stein and Henry Wright, to be an "American garden city" (Stein, 1951, p. 23).

It was during the depression that the planning profession completely adopted the notion and concept of the neighborhood unit. President Hoover's Conference on Home Building and Homeownership (1932) recommended the use of neighborhood unit principles in designing residential areas, and the newly formed Federal Housing Administration (FHA), in its 1936 bulletin on *Planning Neighborhoods for Small Houses*, indicated a preference for neighborhood planning based on the works of Clarence Perry, Clarence Stein, and British neighborhood planning advocate Raymond Unwin (FHA, 1936). Thus, by the advent of World War II, the planning profession, at least, was wedded to the concept of neighborhood and of neighborhood planning.

Planners drew considerable support for their developing notion of the neighborhood as a self-contained urban unit from pioneering work in the field of urban sociology, particularly the work being done at the newly organized University of Chicago school of sociology (Park, Burgess, & McKenzie, 1925). Borrowing from the field of plant ecology, the Chicago sociologists had developed a theory of "human ecology" in which the relationships between individuals, families, groups, and institutions formed into a "natural organization" based on their common location. Natural communities, which were differentiated by segregation into economic and cultural groupings, were viewed as the way individuals "who compose the group [are given] a place and a role in the total organization of city life" (Burgess, 1925b, p. 56).

Furthermore, these sociologists saw the neighborhood as being a mechanism for maintaining urban stability. Burgess (1925a, p. 151), for example, argued that the "village type of neighborhood" and its neighborhood centers acted as a safeguard for youth. Centers or places of congregation outside of the local community, such as the public dance hall, he argued, promoted personal disorganization, delinquency, and promiscuity.

Mobility and change were viewed by the Chicago sociologists as threats to neighborhood stability and the social order. "Where mobility is the greatest and where in consequence primary controls break down completely," Burgess (1925b) argued, "there develop areas of demoralization, of promiscuity, and vice" (p. 59). Not surprisingly, mobility was often equated with the "invasion" of community by undesirable elements, such as foreign races or Negroes (McKenzie, 1925, p. 76).

Thus, although the concept of the neighborhood as a "village" within a larger urban agglomeration of places may have ancient historical antecedents, its planning roots are likely more recent and are associated with the creation of middle- and upper-middle-class residential havens where lifestyles and property values could be protected. This inherently conservative notion was further buttressed by the creation of neighborhood or homeowners associations that served to maintain internal order and to keep out undesirable influences whether they were "incompatible" land uses or "incompatible" individuals or groups.

## ALTERNATIVE NOTIONS OF THE NEIGHBORHOOD

Although the notion of the neighborhood as an urban village still occupies a central place in the thinking of many planners and others interested in neighborhood development and is supported by the early work in urban sociology, it may well be an outmoded and limiting idea. More recent sociologists, such as Gerald Suttles (1972) and Herbert Gans (1991), are critical of the simple Burgess model. Both Suttles and Gans argue that the ecological model of the city is too simplistic and results in a false impression of the city and of urban life.

Suttles (1972) argues that the ecological model is the simplest form of a folk model in which the interactions of urban residents are seen as not extending beyond physical or primordial barriers. He notes that most urban residents do not limit their relationships to those people who live in their local community. Instead, urban residents create and maintain relationships throughout the urban area. These relationships take different forms. Even when there are close ties between individuals in a "neighborhood," the ties usually are found only among immediate neighbors, those found on the face block at what might be considered the subneighborhood level and not among "neighbors" at large.

Gans (1991) argues that the urban ecologists misread the city because they focused too narrowly on their studies of ethnic and immigrant slums that were prevalent at the time. Like Suttles (1972), Gans contends that most urban residents do not limit their contacts to those in close proximity and are more likely to organize around issues relating to social and economic class rather than physical neighborhoods. Even where neighborhoods do exist, the social and economic linkages that the residents create and maintain are not likely to be limited to the neighborhood, even when the neighborhood is highly identifiable and its boundaries distinct.

Gans (1991) identifies five major types of "inner-city" residents: the cosmopolites, the unmarried or childless, the ethnic villagers, the deprived, and the trapped and downwardly mobile. Cosmopolites, who choose to live in places where they can be near to a city's cultural facilities, and the unmarried or childless, who tend to be transient residents, are likely to view the entire urban area as their activity realm. The deprived and the trapped and downwardly mobile[7] have few choices as to where they live, and they tend to be found in areas of the city where there are few if any of the characteristics usually attributed to neighborhoods. Only the "ethnic villagers" tend to isolate themselves and to live in places that can justifiably be called "urban villages." Such ethnic neighborhoods, however, persist only as long as there is a continuing influx of immigrants because older residents die off and their children tend to relocate elsewhere.

Suttles (1972) and Gans (1991) do not deny the existence of neighborhoods, that is, recognizable communities of locations at a somewhat higher level of geographical abstraction. But both argue that neighborhoods are not universal. That is, not every urban place can be thought of as part of a neighborhood. Gans goes further to say that for the most part, there are few differences between most residential areas in a city and the residential areas of its suburbs. Thus, it is incorrect to suggest that there are significant differences between the lifestyles of many urbanites and many suburbanites.

Suttles (1972) defines two higher orders of linkages that exist among urban residents, what he calls the *community of limited liability*, a term previously used by Janowitz (1952), and an *expanded community of limited liability* (Suttles, 1972, pp. 54-64). In each of these higher-order abstractions of geographical space, the linkages between individuals and place weaken, but the fact that people do not live in geographical proximity does not mean that they cannot be connected in a variety of ways, such as by religion, politics, or other common interests. What we have been led to believe are self-contained idealized neighborhoods, Suttles contends, are more often than not the product of "some outsider; a government surveyor, a developer, a Realtor, founding father, booster, or newspaper man" (Suttles, 1972, p. 52).

Jane Jacobs is probably the best-known planner who has been critical of the place-based concept of neighborhood. In her 1961 classic, *The Death and Life of Great American Cities,* Jacobs rejects the notions that neighborhoods have identifiable geographic dimensions and that the ideal neighborhood is one that is a cozy, inward-turned, self-sufficient urban village. City people, Jacobs argues, are mobile and can pick and choose from throughout the entire city and beyond both the goods and services they purchase and their friends and associates as well. She argues that any concept of neighborhood must be fluid, allowing for the linking together of people by interest, association, and purpose. But Jacobs cautions that this does not mean that a city does not need neighborhoods because it is her contention that the quality of a city's neighborhoods—the places where people live, recreate, and undertake their round of daily activities—determines the overall quality of urban life.

Rather than considering neighborhoods simply as places, Jacobs (1961, p. 114) says that we need to think of them as mundane organs of self-government. The success or failure of a neighborhood thus depends on its ability to undertake self-governance. In this context, Jacobs posits three kinds of neighborhoods that she feels are useful for consideration: the city as a whole, street neighborhoods, and districts of large, subcity size (Jacobs, 1961, p. 117). In the context of this discussion, Jacobs is apparently thinking only about large cities because she suggests that districts are areas having a population of 100,000 or more.

Although each of the three kinds of neighborhoods has a different function, they can work together in a complex way to enhance urban life. The bulk of the power in a city is to be found in the city as a whole. It is at this level that decisions are made, policies determined, and resources obtained and distributed. It is also where special interest and pressure groups tend to operate. Jacobs (1961) notes that one of the greatest assets of urban life is the ability of people to come together from various parts of a city to promote a common cause or address a common problem.

At the other end of the scale are street neighborhoods consisting in size of at most a block or two. Street neighborhoods have the ability to effectively organize and maintain order in their small area through the "networks of small-scale, everyday public life and thus of trust and social control" (Jacobs, 1961, p. 119). But they are far removed from the sources of political power and have little control over their ultimate destinies. For any street neighborhood to be fully self-governing, it must have access to the city as a whole and to the power and resources it represents.

The function of districts is thus to act as broker between the street neighborhoods and the city as a whole. "The chief function of a success-ful district is to mediate between the indispensable, but inherently politically powerless, street neighborhoods, and the inherently power-ful city" (Jacobs, 1961, p. 121). Unfortunately, as Jacobs notes, it is at the level of the district[8] that cities most often fail. It is not that districts do not exist in cities, but rather they do not function well, and therefore the street neighborhoods are left isolated from the resources and power they critically need.

Therefore, at least according to Jacobs (1961), effective self-governance at the most local level, the street neighborhoods, depends on the existence of effective linkages to the city as a whole that can only be provided through districts. To be effective, districts must contain people who know and understand the needs and concerns at the street level but who also have "access to the political, the administrative, and the special interest communities of the city as a whole" (Jacobs, 1961, p. 119).[9]

This more flexible notion of neighborhood seems to be more consis-tent with notions of community, for communities of individuals are not necessarily place based. Although many authors, planners, and activ-ists tend to use the words *neighborhood* and *community* interchangeably, the former term is most often associated with specific geography, whereas the latter is most often associated with people living their lives in some common or shared way. A key measure of the strength of a community is not the quality of street maintenance or the aesthetics of a group of buildings but rather the extent to which community members interact with each other, spend time together, rely on each other, cele-brate together, mourn together, and simply talk to each other.[10]

## COMMUNITIES AND THEIR RELATIONSHIPS TO NEIGHBORHOODS

In their work, modern-day planners seem to retain the concept of the neighborhood as an urban village at least in their rhetoric. However, by relying on Euclidean zoning as the primary tool for urban spatial organization, they seem to contradict it. Separating different uses by zones makes it difficult to integrate a variety of activities within a single space. The result is most often something that is claimed to be a neighborhood but in reality is little more than a residentially isolated subdivision with gently curving streets, cul-de-sacs, and possibly a convenience store and gas station situated at the point where the local road running into the subdivision connects with the major arterial road running through the region.

Within cities there still exist places that meet the requirements of Suttles's (1972) defended neighborhoods or Gans's (1991) ethnic villages, although increasingly the most common type of defended neighborhood is the planned middle- and upper-middle-class subdivision with its walls and entrance gates,[11] and what may pass for urban villagers are likely to be low-income minorities, victims of discrimination and racial segregation. For the most part, the typical urban dweller lives in a place that, although it may have a neighborhood name, has few if any distinctive characteristics and differs little from other parts of the city.

Thus, although we all live in place-based locations, which we may refer to either as our neighborhood or our community, the boundaries and meanings of these self-defined places are not likely to be exact. Furthermore, what we consider to be our "neighborhood" may be quite different at different times and for different purposes. We may chat with the woman "down the block" when she walks her dog past our house, vote for the alderman in our ward, send our children to the "local" school, and shop at the local mall. All of these "neighborhoods" make up our community.

We also belong to communities not simply of place but of people who share something in common, such as interests, ethnicity, careers, sexual orientation, and religious beliefs. Communities also form around institutions, such as parents of children attending an elementary school, those men and women who work at the same plant or in the same office, or those who are members of a church.

In reality, we are all members of many communities. And these communities do not exist in isolation. They are interconnected in subtle and intricate ways. And they are not static. They are always changing as members join and leave and as connections to other communities grow or wither.

The above discussion suggests that neighborhood planning is or should be about more than the physical design of local space. And given the more broadly defined notions of neighborhood and community, neighborhood planning certainly should not be about creating isolated, independent, self-sufficient villages in the city. Rather, it should be about building community, and doing this involves identifying not only the local needs but also identifying the ways in which people in neighborhoods link with communities beyond some limited and artificial boundary.

## DOING PLANNING
## AT THE NEIGHBORHOOD LEVEL

Barry Checkoway (1984) has identified two very different approaches to planning at the neighborhood level. He calls the two approaches "subarea planning" and "neighborhood planning." The distinguishing characteristics between the two are far from trivial. Primarily, they differ in terms of who controls the planning process. One is a top-down process; the other is a bottom-up process.

Subarea planning is initiated at the city level and involves the deconcentration of central planning activities to the neighborhood level, whereas neighborhood planning is community based and involves the development of plans and programs by and for community residents themselves. The former is merely, according to Checkoway (1984), "a new form of centralization" (p. 105), and the latter is planning that leads to community empowerment.[12]

## SUBAREA PLANNING

Checkoway (1984) traces the roots of subarea planning to the citizen participation movements of the 1960s and the response by government to claims that it was out of touch with the public. Reacting to this claim, planning departments deconcentrated some planning functions or facilities to subareas of the city or to what were identified as neighborhoods. Checkoway states that subarea planning has the following characteristics: It is "usually initiated by municipal officials . . . may follow steps of rational planning . . . may produce written plans . . . [and] help[s] fulfill minimal requirement for citizen participation in federal funding programs" (Checkoway, 1984, p. 103). As depicted, subarea planning has basically the same characteristics as city planning in general, but it is done on a smaller scale.

Subarea planners are supportive of the aims of citizen participation, but their efforts do not lead to citizen control. Instead, they generally attempt to engage citizens by using procedures that are characteristic with actions of informing, consulting, and placating. That is, they act in ways that are characteristic of what Arnstein (1969) has referred to as the middle rungs of the ladder of citizen participation.

Subarea planners also recognize the importance of community organization, and they will often work with community-based organizations in carrying out their planning activities. But they are cautious about becoming advocates for the community, claiming that their jobs and the planning process itself require objectivity and noninvolvement in political action.[13] As Checkoway (1984) points out, neighborhood planners favor reforms but "oppose measures which would transfer power to local territorial or functional units" (p. 104).

Subarea planning continues to be the approach supported by the American Planning Association (APA) and is the type of planning described in both the 1979 neighborhood planning guide for citizens and planners (Werth & Bryant, 1979) and a more recent 1990 guide (Jones, 1990). In the 1990 guide, Jones defines neighborhood planning as being "nothing very different than the other varieties of plans cities produce. It just deals with a smaller geographic area and rounds out the picture of what forms of planning are needed" (p. 3).

The 1990 APA guide is full of suggestions and methods for involving citizens in the planning process, and it supports the notion of community involvement by identifying a number of neighborhood stakeholder groups, including community organizations, that should be involved in the planning process. But after a generous nod to community involvement, it nevertheless falls back on the rational planning model as the basic process for generating a neighborhood plan and identifies the planner as the technical expert who ultimately produces the plan.

Jones, the guide's author, does not contend that the "objective" planner should be the sole and final arbiter of the neighborhood plan. Instead, he calls for a "democratic" planning process, "a joint undertaking of professional and citizen planner(s)" (Jones, 1990, p. xv). But after saying this, he cautions professional planners, particularly those young planners who might be inclined to develop a "close attachment to their neighborhoods, i.e., the ones with which they are working" (p. xvii), to be careful of potential "minefield(s), strewn with professional and political dangers" (p. xvii). To avoid these minefields, Jones urges planners not to personalize situations, to stay in the middle of controversies, to avoid being in the "middle of the fray," and to convey the neighborhood's position to city officials in a neutral fashion.

## NEIGHBORHOOD PLANNING

Checkoway (1984) clearly prefers the second approach, neighborhood planning, over subarea planning. However, he is not very clear about just what neighborhood planning is. He suggests that like subarea planning, neighborhood planning also has its origins in the 1960s, but he equates it more closely with citizen protest movements than with citizen participation. According to Checkoway, neighborhood organizations in older urban neighborhoods, which "took lessons from their black counterparts" (Checkoway, 1984, p. 105), had by the 1970s become involved in a number of issues that focused on urban decline, including housing revitalization, physical improvement, social services, health and safety, and community empowerment. These issues form the bases of neighborhood planning.

Although Checkoway (1984) notes that some organizations that do neighborhood planning have employed "high visibility tactics," the organizations that he refers to as having done neighborhood planning do not operate in the confrontational mode of community organizing most frequently associated with Saul Alinsky (1972) and his followers. This is because over time, they have found that "it is a strategic mistake to react to crises without an independent agenda of their own" (Checkoway, 1984, pp. 105-106). The creation of an independent agenda or plan, which is a process of political development, is what brings neighborhood planning to be viewed as a process of community development. Thus, for Checkoway, it is the type of organization known as a community development corporation (CDC) that is involved in neighborhood planning.[14]

As will be shown through the case studies of neighborhood planning and development that are presented in Chapters 5 through 8, restricting neighborhood planning to work done by CDCs is too limiting. For the most part, the organizations in these cases are not CDCs, although they sometimes act like CDCs and often use strategies designed to bring about community development. Checkoway (1984) was writing early in the Reagan era and, faced with increasingly conservative ideas about community development, may have been hesitant to suggest that planners link too closely with advocacy organizations. But by failing to recognize the important role of advocacy as it relates to planning, he ignores the entire subfield of advocacy planning that arose in the 1960s and early 1970s as a remedy to traditional top-down approaches to planning.

## ADVOCACY PLANNING

Like the CDC movement, advocacy planning also has its roots in the protest movement of the 1960s, but it is more closely allied with neighborhood advocacy than with community development. Advocacy planning is most often associated with Paul Davidoff (1965), who defined it in his article, "Advocacy and Pluralism in Planning," which originally appeared in the *Journal of the American Institute of Planners*.

Advocacy planning, which flourished briefly during the late 1960s and early 1970s, articulates a rationale for planners to become advocates for neighborhood organizations, whether they are CDCs or community-based or protest-oriented advocacy organizations. In his article on advocacy planning, Davidoff (1965) rejects the notion that planners can act as neutral technicians in the making of plans or that planning involves making choices between technical methods of solutions. Planning objectives, he asserts, are not value neutral, and it is the responsibility of planners not only to identify and articulate the specific values underlying planning prescriptions but also to affirm them. This requires planners to be willing to identify with those objectives in which they believe and to become advocates on their behalf and on behalf of those individuals, groups, and organizations who hold to the same or similar objectives.

Asking why there are no plural plans, Davidoff (1965, p. 333) suggests a model for planning that resembles the legal profession. In his model, planners become advocates for client groups in the public political arena. By making this suggestion, Davidoff argues that planning is a political process and that frequently there are competing perspectives relating to any planning issue. The planner as an advocate would believe in and support the perspective of the client and would use his or her professional skills to be a proponent for this perspective.

Breitbart (1974) has contrasted the nature of traditional conventional planning against that of advocacy planning. She identifies four basic differences:

1. Although in the traditional mode, planners work as technical experts outside of the realm of politics, advocacy planners recognize that every decision has a political meaning attached to it and do not approach resource allocation problems as though they are "value free."

2. Although traditional planners see themselves as working to evaluate interests of the "general populace" and arriving at goals and objectives that address these interests, advocate planners believe that there can be no "general interest" and that they must work to develop "plans" that more adequately reflect the interests of the less powerful in society and to make planning issues understandable and accessible to all elements of society.

3. Although traditional planners assume that through their expertise, they should be entrusted to allocate resources to meet the needs of the "general populace," advocacy planners believe they should not "dictate" to

the people but instead should act as advocates on behalf of clients, so that they can have a voice in the planning process.

4. Although traditional plan making strives to be "objective" and is characterized as a process of objective fact finding, advocacy planners implement their own values by representing clients of similar views and are able to assist underrepresented groups to address injustices brought on them by the implementation of "objective" public policies (Breitbart, 1974, p. 44).

With whom are advocacy planners to work? Davidoff (1965) identifies three types of organizations for which the advocacy model seemed to him to be appropriate: political parties, special interest groups, and ad hoc protest associations. He holds out little hope for involving political parties in any kind of planning process and further suggests that special interest groups may find it difficult to directly engage in planning in the public arena for fear of alienating one or more segments of their support. Davidoff concludes that it is groups focused on one or only a few issues that are most likely to accept the notion of advocacy planning. He argues that the most realistic place to expect advocacy planning to occur is at the neighborhood level. He also suggests that advocacy planning efforts should be likened to legal aid and thus funded by either foundations or the federal government.

Advocacy planning flourished for a brief period in the late 1960s, when there was a considerable commitment by government at all levels to address problems of the cities and the plight of the urban poor. Some planning agencies even added advocacy planners to their staff. When I was in my first teaching position at Bowling Green State University in Ohio, I learned that two advocate planners were employed by the Toledo Lucas County (Ohio) Planning Commission. One worked in a lower-income ethnic community on Toledo's "east side," and the other worked with residents at the fringe of the built-up portion of the metropolitan area who were facing the onslaught of urban sprawl. Wanting to find out more about these advocate planners and the work they were doing, I called the Planning Commission office, but it took considerable effort to find anyone there who would or could confirm that in fact these two individuals were employed by the agency. Even though the two received their paychecks from the Planning Commission, their contacts and relationships with it were quite minimal, bordering on being nonexistent, and their work on behalf of their client groups sometimes brought them into conflict with their own employer.[15]

As the 1970s progressed, interest in and support for advocacy planning declined as governments at all levels reduced their commitment to addressing problems of poverty and the inner city, and many cities were faced with financial crises. Advocacy planning positions within government, which had been daring and controversial from the outset, were eliminated, although many planning departments still tried to maintain a focus on neighborhood planning.[16] Foundations, which Davidoff (1965) had thought would be sources of funding for advocacy planning, never materialized as supporters. Foundations were generally more interested in funding direct community organizing. As time went on, foundation funding began to focus on the funding of community development corporations that engaged in the "bricks and mortar" of neighborhood revitalization rather than supporting neighborhood planning and advocacy efforts.[17]

## EQUITY PLANNING

It has always been difficult for planning agencies to do neighborhood planning in ways such that the concerns and issues of neighborhood residents are given primary attention. Alan Jacobs (1980), writing of the period from 1967 to 1974, during which he was the director of city planning in San Francisco, chronicles how hard it was to balance the interests of various constituencies. He says that he had to try to overcome citizen distrust, create a rapport with the mayor and the board of supervisors, address the bulldozer approach to city problems that characterized the federal approach to urban renewal, and respond to the interests of the city's neighborhood organizations. Although he was able to establish a planning presence that was "reasonably accessible and responsive" and that led people "to feel they had a place to go, a place to be heard," the political pressures of attempting to be a people's planner and to meet the demands of the political process ultimately limited what he was able to accomplish.

Only a few planners and planning agencies have managed to successfully do bottom-up neighborhood-focused planning from within the government. Advocacy planning, when it is carried out as part of a planning agency's regular activities, has come to be called "equity planning" because its goals can be generally thought of as creating social equity and redistribution (Metzger, 1996). The most well known and extensive application of equity planning has been associated with the work of Norm Krumholz and his colleagues in Cleveland during the 1970s.

Krumholz retained his position as director of the Cleveland City Planning Commission from 1969 to 1979 through three different city administrations (Krumholz, 1982; Krumholz & Forester, 1990). In a retrospective 1982 article, Krumholz asserts that he and his coworkers were able to take advocacy planning from a theoretical standpoint to a "tangible effort undertaken within the system and directed toward, and achieving, real ends" (p. 163). This was accomplished, according to Metzger (1996), by paying less attention to growth and development, particularly downtown development, and by paying more attention to issues of poverty, neighborhood abandonment and disinvestment, and inequitable service delivery. The primary goal of equity planning in Cleveland, according to the Cleveland Policy Planning Report (Cleveland City Planning Commission, 1975), was to give priority attention to the task of promoting a wider range of choices for those who had few, if any, options and to do this within a context of limited resources and existing pervasive inequalities.

There have been other equity planning efforts in other cities. For the most part, these have been associated with a specific progressive city administration, and the equity planning ended when that administration left office.[18] For example, during the Harold Washington administration in Chicago, planning efforts under the direction of Rob Mier and Elizabeth Hollander (see Mier, 1993; Clavel & Wiewel, 1991) were carried out from an equity perspective, although as will be pointed out in Chapter 7, not even Rob Mier's deep personal commitment to social justice and equity could keep him from occasionally bending to political pressure and supporting development that was directly detrimental to neighborhood residents. Despite many equity planning successes in Chicago, planning quickly reverted to business as usual in the city after the death of Mayor Washington in 1987.[19]

Progressive neighborhood-focused planning—whether it is done at the neighborhood level carried out by community organizations or community development corporations, by advocacy planners working with communities, or by equity planners from within city govern-ment—is closely associated with the notion of community or neigh-borhood empowerment. But like *neighborhood, empowerment* is a term that, although often used, is rarely defined. Thus, before proceeding to present the set of case studies involving neighborhood planning in Chapters 5 through 8, I will focus in the next chapter on both the notion of empowerment and how, at the community or neighborhood level, it is addressed.

## NOTES

1. Several years ago, when I was preparing a lecture on neighborhoods, I reviewed several books and articles on the subject. Although all of them used the word *neighborhood,* the term was never defined.

2. Mumford (1961) points out that the neighborhood unit was often identified with the church parish and would derive its name from the parish church. In heavily Catholic cities, such as Chicago, this custom continues.

3. During the Renaissance, Bristol was England's third most populous urban center and its second most active port (Sacks, 1989). Thus, it qualifies as a major urban place in 15th- and 16th-century England.

4. One of the reasons that Pullman chose to build a model town, locating it far from what at the time was the built-up area of Chicago, was that he could have more control over the workers, provide them with a wholesome environment, and keep them away from bad influences, such as union organizers. His attempts failed, however, as the city rapidly expanded southward, his workers found ways to get to the bars in the nearby "Schlitz district," and his workers became dissatisfied with the company's heavy hand. The internationally famous Pullman strike of 1894 effectively ended the notion of Pull-man as a worker's paradise.

5. Until the financial reforms and the rise of the "building and loan" industry during the depression, it was very difficult for even middle-income households to be homeown-ers. Banks required high down payments, and loans were generally extended for no more than 5 years (Worley, 1990).

6. Nichols used the country clubs and the parkways between his roads as his parks. In two of Nichols's subdivisions, Crestwood and Wornall Manor, interior parks were ac-tually planned in the middle of the blocks so that they could only be reached from the backyards of the houses. These "parks" were clearly intended as recreational areas only for those families that lived on the block (Worley, 1990).

7. Gans originally wrote this discussion of urban and suburban life in 1962. In his 1991 book, he suggests that his division of urban types may be somewhat time bound. The names he gave each group certainly need changing. Today, the unmarried and child-less would be called "yuppies" and "dinks" (double income, no kids). The "trapped" are now more commonly referred to as the "underclass."

8. For a city such as Chicago, we might think of districts as being wards (there are 50 in the city) or the officially designated community areas (there are 77 community areas).

9. Jacobs (1961) has been criticized for idealizing the street neighborhood. Her model of the ideal street neighborhood was her own New York Hudson Street neighborhood, where she was impressed by the level of street activity and what she called the "intricate ballet" that took place daily on the sidewalk outside of her home. People knowledgeable with urban life recognize that not all street neighborhoods have so many interlocking relationships and activities. In some neighborhoods, very few residents know each other. In others, the "intricate ballet" may be performed primarily by drug dealers, prostitutes, and their clients.

10. My thinking on the differences between neighborhood and community has been influenced significantly by a series of discussions with colleagues while performing a community development needs assessment for the Chicago United Way from 1997 to 1998. The material in this and in the following paragraphs is drawn heavily from these discussions.

11. A few years ago, on one of my noontime jogs from the campus of the University of Illinois at Chicago, I decided to explore the then new cluster of town homes in South Dearborn Park, a redeveloping area located south of Chicago's Loop. I entered the area from the east, planning to run through it and exit at its west end. After failing to find any open routes to the west, I reluctantly ran back to the same place I had entered, realizing it was the only way to get into or out of this development that occupied several city blocks.

12. Checkoway (1984) does not actually use the word *empowerment*. As will be pointed out in Chapter 3, the word *empowerment*, as applied to community development, is of rather recent origin.

13. To support this assertion, Checkoway (1984) points to the 1979 American Planning Association guide to neighborhood planning, which warns planners that

> it would be unwise to adopt the Alinsky approach. Not only do (planners who adopt the Alinsky approach) jeopardize their jobs by involving their programs and agencies in controversy, but, more important, they jeopardize the nonpartisan role as link between the neighborhood and city agencies. (Werth & Bryant, 1979, p. 16)

14. That Checkoway (1984) links neighborhood planning with community development corporations (CDCs) becomes particularly clear near the end of his article when he references Mayer and Blake (1981) as having done an analysis of neighborhood planning organizations. In their analysis, Mayer and Blake do not use the term *neighborhood planning organizations* but instead call the organizations they studied *neighborhood development organizations,* which is a term generally considered synonymous with community development corporations.

15. I eventually made contact with both advocate planners, although the person working in the east Toledo community showed little interest in speaking to me about what he was doing. I did come to know the other advocate planner, Marcy Kaptur, and my own first planning effort working in an advocacy mode was on a follow-up to a project that she had been involved in working with residents of Springfield Township, an area just west of Toledo. Kaptur left Toledo in the mid-1970s and moved to Chicago, where she became an advocate planner for the group that developed a widely acclaimed "Community 21" plan for the "west-town" area of the city. She later returned to Toledo and subsequently was elected to represent the Toledo area in the U.S. Congress. Today, she is one of the few members of Congress with a planning background and undoubtedly the only one who has worked as an advocate planner.

16. In 1977, I got the opportunity to spend a year working essentially as an advocate planner for the Toledo Metropolitan Mission, a branch of the Toledo Council of Churches. The position was funded by a National Science Foundation (NSF) program called Public Service Science Residencies. The NSF somewhat naively presumed that the recipients of the residencies would "interpret" science to agencies and organizations that lacked scientific expertise to improve the groups' abilities to participate in the policymaking process. Most of the recipients, however, were primarily interested in advocacy work of one kind or another. The program was short-lived and was canceled after a second group of residencies was funded.

17. For some time, advocacy planning has been a topic planning students read about, usually being assigned Davidoff's (1965) seminal article in their introductory planning theory class. But in most planning curricula, students get little or no experience in doing advocacy planning. Occasionally, when they encounter a faculty member who participates in some kind of action-oriented research or practice, students get a chance to do some hands-on advocacy and neighborhood planning work. A few examples include planning students at the University of Illinois at Urbana-Champaign who may get the opportunity to do "real community-based planning in East St. Louis, Illinois" (Reardon, 1997, p. 233) under the tutelage of Ken Reardon; graduate assistants at the University of Illinois at Chicago, who work with either the Voorhees Center for Neighborhood and Community Improvement or the Center for Urban Economic Development; and students in urban affairs at the University of Delaware who work for CDCs, local governments, and other not-for-profit organizations.

On graduation, a few planning students take jobs with neighborhood organizations or community development corporations and thus become neighborhood advocates. However, the amount of time these former planning students devote to actual planning may be quite limited. For the most part, however, within the planning profession, advocacy planning remains something about which students read and are presented with as an idea whose time has passed.

18. In addition to Chicago, mentioned in the remainder of the paragraph, other notable equity planning efforts have included the work of planners such as Peter Drier in Boston and Derek Shearer in Santa Monica (see Metzger, 1996, for a thorough review of equity planning).

19. Some of my colleagues might disagree with me on this point, arguing that although the current Daly administration is far different from that of Harold Washington, some changes have been made. Although I am willing to agree that planning in Chicago in the 1990s is certainly different than it was in the 1970s, change has been much more rhetorical than substantive.

# CHAPTER 3

# COMMUNITY EMPOWERMENT, ORGANIZATION, AND DEVELOPMENT

The previous chapter focused, for the most part, on planners, the planning process, and the role planning plays relative to neighborhoods and community development. In this chapter, the focus shifts to the neighborhood or community. The purpose of this chapter is to explore how neighborhood people can and do act to bring about neighborhood development, particularly to explore how neighborhood people become organized to take positive action. After placing neighborhood planning and community organization in the context of the modern American city in the next chapter, the blending of planners and their work and neighborhood residents and their organizations is then explored through the case studies in Chapters 5 through 8.

Based on what was discussed in the previous chapter, we are led to believe that neighborhoods are not isolated places where people live out their lives unaffected by what is happening outside of some arbitrary physical space. Indeed, neighborhoods are more than physical places, and rarely do they have distinct boundaries. Neighborhoods, or at least the people and institutions that constitute a neighborhood, are involved in a myriad of daily interactions, some between entities internal to the neighborhood, but many others involving individuals, groups, and institutions external to the place we might call the neighborhood. It is the frequency, strength, and quality of all these interactions, both internal and external, that determine the stability and vitality of the neighborhood. It is what makes a neighborhood a community.

An important goal of neighborhood planning should be the maintenance or creation of conditions and situations that help to maximize, when and wherever possible, the internal and external linkages experienced by a neighborhood's residents. Note that this suggests that physical planning and the arrangement of amenities within a neighborhood alone cannot make for a good neighborhood. Having a neighborhood park that is designed to be used by the children of the neighborhood is of no use if drug dealers appropriate the park space and parents are afraid to send their children to the park for play. Of course, a well-designed park is better than no park, but the mere existence of a park is insufficient to create community.

Any place where people live can be thought of as a neighborhood. Good neighborhoods happen, however, when people strive to turn a place into a community. Neighborhood planning therefore should primarily be about helping people create and build community. Sometimes this may mean creating physical structures that support a local place-based community. But even when this is being done, the purpose is not to isolate the neighborhood by providing it with local resources. Creating local resources can provide people with the means to access additional resources that exist beyond the local area and to create linkages with other neighborhoods and with other individuals and organizations throughout the city. It is the combination of maximizing internal resources and the ability to access necessary external resources that results in community empowerment and successful communities.

## WHAT IS EMPOWERMENT,
## AND WHY IS EVERYBODY FOR IT?[1]

In this chapter, I address the notion of community empowerment and define its meaning with respect to neighborhood development. *Empowerment,* like neighborhood and community, is a term often used but seldom defined. The lack of a clear definition of empowerment is probably why people with very different political agendas can be for it. But different meanings of empowerment will and do result in different ideas as to how it is achieved. Thus, before we can move ahead and talk about neighborhood development, which almost always involves some notion of empowerment, we must first look at what it means to be empowered.

Once we have a clear definition and understanding of what empowerment is and how it relates to neighborhood and community, we can review the ways in which communities can become organized to achieve empowerment. So the second half of this chapter will be devoted to a discussion of the various types of community organizations, community organizing, and the different ways in which organizations attempt to affect positive neighborhood change.

The first time I remember hearing the word *empowerment* was in 1984, during a presentation by Washington, D.C., public housing activist Kimi Gray. Gray had come to Chicago, along with conservative policy analyst Robert Woodson, to meet with tenants at Leclaire Courts in Chicago (the Leclaire Courts story is presented in Chapter 7). As she spoke to the Leclaire residents about how tenants had organized to take over the management of the Kenilworth-Parkside public housing development, she repeatedly stated that this process had led to resident empowerment. Gaining control of the management of a housing development, at least as Gray described it, seemed to be synonymous with resident empowerment.

In the next few months as I worked on the resident management issue, I began to hear the word *empowerment* used more and more. Robert Woodson, who had accompanied Gray and was the head of the National Center for Neighborhood Enterprise (NCNE), a conservative think- tank that had been spun off from the Heritage Foundation, used the word a lot. But I also heard it used by staff at the Metropolitan Planning Council (MPC), a liberal civic organization in Chicago. The MPC staffers asserted that resident empowerment would result from shifting Chicago's high-rise public housing from Chicago Housing Authority management to resident management. Activists and academics further to the left of MPC were also using the term. In fact, I myself began to use it, although I really was not sure what it meant. Just about everybody, it seems, was claiming that empowerment was a positive outcome of turning over the responsibility of managing public housing to the residents.

Implied in all this talk about empowerment was the assertion that empowered tenants taking control would result in better public housing (i.e., better public housing neighborhoods). But how better public housing actually would happen and whether empowerment was the means to it or simply a process that eventually resulted in better housing and thus better neighborhoods was unclear.

In 1987, after I had more or less completed my work with the residents at Leclaire Courts, I was fortunate to receive a small grant from the MacArthur Foundation for the purpose of reflecting on what had and had not been accomplished by converting Leclaire to resident management. One of the things that I was particularly interested in investigating was the notion of empowerment.

I already knew that resident management did not originate in the 1980s, and I soon came to discover that it had not always been so closely linked to empowerment. Resident management had first appeared in both Boston and St. Louis in the early 1970s and, in the mid-1970s, had been expanded to several other cities by means of a jointly funded U.S. Department of Housing and Urban Development and the Ford Foundation demonstration project. As Chandler (1991) has noted, the term *empowerment* was neither used during these early efforts, nor did it even seem to be an implied goal of what was then referred to as tenant management. Instead, the efforts during the demonstration project of involving tenants in the management of their housing had more modest goals of being a "means to decentralize some housing authority responsibilities and create a bit more stability in the resident population" (Chandler, 1991, p. 137). If this was empowerment, it certainly was not what Woodson and others seemed to have in mind.

As I explored the notion of empowerment further by talking to various people involved in the resident management movement and reading the literature that was being produced, I came to realize that the word had a variety of meanings, and its specific meaning at any given time depended on who was using it. This was why all sorts of people, coming from all sorts of political philosophies, could agree that empowerment was an important concept and that it was what resident management was about. In reality, however, there was little agreement because the meaning of the word, as used by any one individual, was closely tied to that person's political philosophy. After a while, I came to identify three basic and different meanings of the term and to associate the meanings with three basic political perspectives: conservative, liberal, and progressive.

Political conservatives associate empowerment with ownership. They value individualism and emphasize the role of private property in maintaining the social order (Perin, 1977). To them, becoming empowered means to become free from the dependency of bureaucratic control and to become an owner and controller of property.

This means that with respect to empowering tenants through resident management, the ultimate goal should be tenant ownership of the housing development. Bringing people together by creating a community organization is the way or the means for people to work toward the end result, individual ownership. Thus, communities and community organizations are public means to private ends.

Liberals have a whole different notion of empowerment. To them, one becomes empowered through access to government. Thus, tenant management means the sharing of power through tenant participation. From this perspective, it is unnecessary for tenants to become owners. In fact, they do not even have to be fully in control of the housing development. By becoming involved in or responsible for some of the management tasks, tenants are given a voice in making decisions about the current state of the development and in planning for its future. Community organization as a tool used in resident management, in the sense that liberals use it, is the necessary mechanism for bringing about tenant participation.

Progressives, or those on the political left, equate empowerment with the notion of community control. Control is obtained by tenants coming together to take power away from those who have oppressed them.

Community organizing and community organizations are the means for becoming powerful and for taking control. Empowerment, therefore, is the result of successful community organizing and organization.

Although conservatives equate empowerment with personal control and economic independence, liberals and progressives equate it with political control that occurs when "individuals organize into a strong group that applies pressure and becomes recognized as a force in the community" (Bratt, 1991, p. 174). Thus, for liberals and progressives, empowerment is an end or goal, and the means for reaching the goal are community organizing and community organizations.[2] In every case that I was aware of, public housing residents in the resident management movement had been able to wrest control away from their housing authorities by forming a community organization that spoke and acted on behalf of the individual tenants.

In some situations, a resident management effort did not result in the residents actually taking control of their development but instead merely resulted in the transfer of some or even, in some cases, a significant number of management tasks to a tenant corporation. In these situations, even though there had been resident organizing, it had not led to empowerment. Empowerment, it seemed, happened only when tenants were given the power to make decisions about the future of their developments and the resources needed to implement them. And this happened when tenants were given control over the budget for their housing developments.

What I concluded through my study of the resident management movement was that political empowerment or control, brought about through community organization, seems to be a necessary, though, as will be discussed later, not a sufficient requirement to bring about neighborhood-driven community development. The process of becoming empowered and the role planning does or can play in this process is what will be explored in the rest of this chapter, in the case studies that follow, and in the analysis presented in Chapter 9.[3]

## PLANNING AND CITIZEN PARTICIPATION

The belief that planners, using the rational planning model, could articulate goals for and speak on behalf of a community without the direct involvement of the community came under critical scrutiny as the urban renewal programs of the 1950s and 1960s resulted in the destruction of community after community. People in areas targeted for urban renewal found it hard to agree with the pronouncements of urban planners and to see how they would benefit from being dislocated and having their communities destroyed (see Gans, 1962).

Community activists fought to gain at least some control over the urban renewal process, and over time both federal and local governments were forced to include citizens in the planning process. What started out as a vague and undefined requirement that communities and their residents be informed and given the opportunity to participate in developing and administering urban renewal programs in the Housing Act of 1954 eventually became a requirement for maximum feasible citizen participation in the Model Cities Act of 1966. Furthermore, the wording of the Housing and Community Development Act of 1974 appeared to expand the process of citizen involvement to an even greater degree, by turning decision making about community revitalization over to local governments and by involving community organizations in the implementation of revitalization programs. However, even though the role of citizen involvement seemed to expand, just how citizens or their organizations would be able to use the process of citizen participation to really take control of their communities was not always clear, and what was often touted as citizen participation in reality amounted to little or no real citizen involvement, let alone citizen and community control.

The classic essay on citizen participation in governmental policy-making was and still is an article written by Sherry Arnstein (1969) in the *Journal of the American Institute of Planners.* Arnstein lays out a typology of eight levels of citizen participation in the form of a "ladder" for which the lowest rungs offer options that result in little or no real citizen control of the planning process and the highest rungs offer complete citizen control. Arnstein's basic argument is that the lower rungs of the ladder are shams and procedures used by those with power to avoid citizen input and that real citizen participation occurs only when citizens are given power and control over decisions regarding how the benefits of society are to be shared.

At the lowest levels of the ladder are *manipulation* and *therapy*, two rungs referred to by Arnstein (1969, pp. 218-219) as nonparticipation.

Citizens are *manipulated* when their representatives are placed on rubber-stamp advisory committees or boards that have little or no power.[4] Often, members of these committees or boards are not provided with information needed to make informed decisions and are merely asked to vote on decisions that have already been made. *Therapy* masks as citizen participation by attempting to recast a systemic problem into an individual problem. Citizens are led to believe that they have become involved, but their involvement is not directed toward solving the real problem. Most often, citizens are told that they themselves are the problem and their involvement should be directed toward improving themselves or their neighbors.[5]

Moving up the ladder, we reach the next three rungs—*informing, consultation,* and *placation*—which Arnstein (1969, p. 217) groups together as degrees of tokenism. In general, these are one-way processes in which citizens are made aware of what is happening and may even be asked to provide input, but their interests and wishes may not really be considered. Even when attitude surveys, neighborhood meetings, and public hearings are used, citizen participation can be a charade, as long as no serious efforts are made to incorporate the will of the community into the process.

At the top of Arnstein's (1969) ladder are three rungs—*partnership, delegated power,* and *citizen control*—all of which involve a real transfer of power from government to the citizens. In partnership arrangements, power is shared, but in delegated power arrangements, citizens have full control over some but not all aspects of a process or program. Full citizen control "guarantees that participants or residents can govern a program or an institution, be in full charge of policy and managerial aspects, and be able to negotiate the conditions under which 'outsiders' may change them" (Arnstein, 1969, p. 223).

The empowerment of public housing communities through resident management, as discussed earlier and in greater detail in Chapter 7, involves action at the top three rungs of Arnstein's (1969) ladder. How citizen involvement results in empowerment, however, depends on one's political perspective. From a conservative point of view, citizen participation and subsequent citizen control are intended to eliminate, as much as possible, the involvement of government and to maximize individual control and ownership. Liberals' belief in government makes them hesitant to advocate for participation at the highest rung of the ladder and instead argue for *delegated power* and a sharing of power, as well as control between government and community. The rhetoric of the left or progressive activists would appear to suggest that only the highest rung of the ladder, *citizen control*, results in true empowerment, but in reality most progressives are willing to accept arrangements that are somewhat less than full citizen control. Such full control is unachievable in public housing, for example, as long as local housing authorities remain the owners of the buildings and land, and residents and their organizations are required to abide by federal laws and regulations.[6]

## COMMUNITY ORGANIZATIONS
## AND COMMUNITY ORGANIZING

The discussion of empowerment so far has focused almost exclusively on public housing neighborhoods and the use of resident management as the tool of empowerment. Empowerment, resulting from citizen participation that leads to citizen control and citizen power, however, has broader applications and can be viewed as the ultimate goal of citizen action in any kind of neighborhood. In most instances, the mechanism by which empowerment and control are achieved has been the community organization. Therefore, I now turn my attention to community organizations, their structures and intents, and how they act to assert power for control and community improvement.

As noted in Chapter 2, Jane Jacobs (1961) has identified two poles of urban life—the city as a whole, where most of the power resides, and the street neighborhoods, where the common citizen experiences everyday life. Jacobs argued that if street neighborhoods have some measure of control over their existence, they must be able to gain access to power. They can only accomplish this by working through "districts" that act as mediators between the "inherently politically powerless street neighborhoods, and the inherently powerful city" (Jacobs, 1961, p. 121). Districts function through organizations representing several geographically related street neighborhoods and advocating for them at the city level. These organizations are what are commonly called community organizations.

Citizen organizations or associations are not something new. They have been a fixture of American life since at least the early 19th century (de Tocqueville, 1835/1945). One explanation for their rather ubiquitous existence is that they represent a uniquely American response to the expansion of democratic ideals and concern for the common good that swept through the Western world during the 19th century. Although in Europe there existed at the time well-established and strong central governments, the fledging U.S. government was relatively small and weak. Thus, the expansion of democratic ideals throughout Europe could be accomplished through governmental action, but in the United States, it was left to voluntary citizen-supported organizations.

Take, for example, the notion that all citizens should have access to some level of education. In Europe, the response was the development of national systems of education supported by national standards. In the United States, education became a local issue, handled by local school boards that set standards, determined curricula, and hired and fired teachers. Similar action was taken with respect to the extension of health care throughout society. In the United States, hospitals were built by religious associations or other charitable societies rather than by the government.

The roots of the modern neighborhood organization are to be found both in upper-class residential developments and in lower-class slums. In wealthy communities, homeowners, at the urging and insistence of real estate developers, formed organizations to ensure the continuing exclusivity of their neighborhoods and to protect their property values. In poor communities, social reformers worked to bring people together to advocate for a better life, especially better housing and improved social conditions.

The neighborhood association, a form of self-governance among homeowners, first appeared in England in the mid-18th century and then in the United States in the mid-19th century (Worley, 1990). A creation of the developers of upper-class subdivisions, homeowners' associations were ways to minimize governmental intrusion, ensure that the rules of the subdivision would be maintained, and help guarantee that the value of the property in developed communities would remain high so that the developer could continue to sell his or her property. The homeowners association became both a neighborhood booster and a neighborhood watchdog, holding community-building events and doggedly enforcing rules regarding the appearances of homes and lawns and proper social behavior.

In many instances, the associations acted as local governments, providing such services as street sweeping, water, and even security. Today, homeowners associations remain commonplace in new suburban subdivision developments, and they are also frequently found in many city neighborhoods wherever the percentage of owner-occupied homes is high.

Community advocacy organizing and related community advocacy organizations, however, are more a product of the social reform movement of the late 19th and early 20th centuries and are usually associated with poor and immigrant neighborhoods. As the problems of urban slums became increasingly apparent and more serious in the latter part of the 19th century, Americans looked to the British settlement house movement as one way of addressing the needs of the poor. Settlement houses sprung up in many U.S. cities, following the establishment of Hull House in Chicago by Jane Addams. Although the main job of the settlement house was to deal with the everyday needs of the urban poor and to integrate working-class immigrants into American society, settlement house workers began to seek out the causes of poverty and deprivation and, in doing so, became pioneers in social planning and community organizing. The settlement house thus became a source of community organizing and of the modern community organization (Austin & Betten, 1990).

One of the most aggressive examples of an early community organizing effort was the short-lived Cincinnati Unit Experiment established in 1917 in Cincinnati's Mohawk Brighten neighborhood (Austin & Betten, 1990). The Social Unit was an attempt to organize a single neighborhood through block-by-block organizing "to give its people partial control over their immediate social and economic life and to provide a way for residents themselves to deal with social problems" (Austin & Betten, 1990, p. 36). Block workers, women drawn from the community, were the focal point of the experiment. These women were "educators and interpreters, interpreting Unit programs to the people and relaying the people's desires and needs to the Unit administration" (Austin & Betten, 1990, p. 39). With its base of block organizations, the experiment carried out its neighborhood level work through a Council of Neighbors, which was a district-level entity in which each of the blocks had a representative. Each Council of Neighbors, in turn, elected a representative to an overall Citizens Council, which was the governing body of the experiment. Thus, each individual block or street neighborhood, relatively powerless by itself, was linked to the more powerful council, which could advocate on behalf of individual residents.

The confrontational tactics of social activist Saul Alinsky form the bases for much of modern-day advocacy organizing and are the primary tools used by advocacy organizers to help bring power to relatively powerless groups (Ross & Levine, 1996). Alinsky's organizing techniques were rooted in the labor movement, particularly the Congress of Industrial Organizations (CIO) of the 1930s (Austin & Betten, 1990, p. 152). Beginning in 1939 with the Back of the Yards Council, Alinsky used the tools of controlled conflict as a way for community residents to confront those groups and individuals who possessed economic power, political power, or both. The Industrial Areas Foundation (IAF), the organization Alinksy founded to train organizers and to carry out organizing campaigns, espoused an organizing strategy based on direct confrontation of authority as a way of demonstrating community power and as a means of achieving neighborhood goals.

The community-based organizations that Alinsky was instrumental in founding were the agencies that linked the powerless people in the neighborhoods and the powerful people that ran the city. Thus, they acted in the manner of Jacobs's districts (Jacobs, 1961). Alinsky understood, as will be discussed in the next chapter (see Bowden & Kreinberg, 1981), that cities were about power and not about neighborhoods. His most significant contribution to the neighborhood movement was his use of an upfront, in-your-face style of confrontation as the primary tool for getting the attention of powerful people (see, e.g., Alinsky, 1972).

By using confrontational actions, often outrageous and obnoxious, Alinsky led his organizations in challenging the urban power structure and in making sure that neighborhood concerns were addressed.[7] Alinsky and those who have followed him are not particularly interested in actually taking control of the city but instead have as their primary goal the forcing of those who have the power to do what the neighborhood wants them to do.[8]

Many subsequent community organizers have used Alinsky's organizing style but have come to believe that simply getting public officials and others with power to recognize, listen, and respond to the concerns of the community may not be sufficient to bring about real neighborhood change. These organizers recognize that there are often political, social, and economic inequities that define the relationships between those in power and those in the community, and without at least some structural economic and social changes, the needs of communities will not be met.

Gale Cincotta, founder of the National Training and Information Center and its organizing arm, National People's Action (NPA), is probably the best-known community activist to use Alinsky-style tactics to bring about changes in federal and state laws that benefit neighborhoods and neighborhood revitalization efforts. To address what Cincotta and others have identified as a "cycle of decline" that has occurred in many cities, the NPA-affiliated organizations successfully lobbied the U.S. Congress for passage of the federal Home Mortgage Disclosure Act (HMDA) in 1975 and the Community Reinvestment Act (CRA) in 1977.

The primary cause for the cycle of decline was and is redlining, the refusal of banks and savings and loans institutions to make home mortgage and home improvement loans in neighborhoods perceived by lenders to be in decline, and by abuses of the Federal Housing Administration (FHA) loan program. Under the FHA's program, both lenders and real estate agents encouraged home ownership by unqualified families, which in turn led to loan defaults, foreclosures, and subsequent abandonments (Naparstek & Cincotta, 1976). Analyses of bank lending patterns, using HMDA data, and campaigns mounted to enforce the CRA have become major tools used by neighborhood organizations and community development corporations to spur neighborhood revitalization through the influx of investment capital into heretofore capital-starved areas of cities.[9]

There is an alternative to the type of community organizing of Naparstek and Cincotta (1976). Its different goals and style have been articulated most effectively by policy analyst Milton Kotler (1969) in his book *Neighborhood Government: The Local Foundations of Political Life*. Like others, including Jacobs, Alinsky, and Cincotta, Kotler felt that city government was too far removed from the hopes and concerns of the common people. However, he argued that this had come to be simply because cities had grown too large to be self-governed.[10] His solution, therefore, was to bring government to the people by decentralizing its powers and functions. For Kotler, community organizing and community organization were vehicles for bringing about local governance.

Kotler's (1969) ideas were quite popular in the late 1960s and early 1970s, a time when city, state, and federal governments seemed out of touch with the people. His work appeared at the same time that environmentalists, who often were neighborhood activists as well, were becoming acquainted with the "small is beautiful movement" led by British economist E. F. Schumacher (1973). Activists David Morris and Carl Hess, in their book *Neighborhood Power* (1975), argued that it was possible to create a self-sustaining, environmentally sound, and self-governing community in which community interests would come first. The National Association of Neighborhoods, a coalition of neighborhood organizations, which was based on Kotler's ideas, formed during the 1970s and remained an important force in the neighborhood movement well into the 1980s.[11]

In Chicago, political scientist Dick Simpson, who was alderman of the city's 44th Ward during the 1970s, instituted a "ward assembly" based on the local governance concept (Salem, 1993). Simpson's 44th Ward Assembly was made up of representatives from the various local organizations in the ward. Through frequent assembly meetings, Simpson would keep the community representatives informed of issues coming before the Chicago City Council and in return would listen to what the representatives had to say about these and other issues. Most important, before casting his vote in the council, Simpson would first get the approval of the assembly. His intended vote could be overturned by a two-thirds vote of the assembly.

The movement for local governance continues today in Chicago in the school reform movement (Vander Weele, 1994). Sensing a need to have stronger community input into the way schools in Chicago were being managed, school reform activists successfully lobbied for state legislation, which, among other things, established local school councils at each elementary school and high school. The councils, which have certain powers of policymaking, are made up of teachers, parents, and community residents. Although the councils have been and continue to be controversial, and it is not clear how much reform they have been able to accomplish, they seem to have become a permanent fixture of the Chicago public school system.

## COMMUNITY DEVELOPMENT CORPORATIONS
## AND COMMUNITY DEVELOPMENT

Community development corporations (CDCs), which contain elements of both community organizing and economic and physical development, represent yet another organizational approach to neighborhood control and empowerment. CDCs are locally based quasi-capitalist organizations that attempt to stimulate and carry out community development efforts in neglected communities where the private market has failed to do so (Rubin, 1993). The most common activity undertaken by CDCs is the renovation or construction and subsequent management of low- or moderate-income housing. Some CDCs, however, pursue goals of economic development, either carrying out business or industrial activities or encouraging and assisting others to do so. A few CDCs are comprehensive development organizations and are involved in housing, economic development, social service delivery, and even the arts. The argument made in support of CDCs is that they promote community empowerment through the successful application of business skills tempered with social awareness (Rubin, 1994). Their dual focus on economic and social empowerment, however, raises questions as to whether CDCs are a complementary addition to advocacy-based neighborhood efforts or conflict with them.

Community development corporations first appeared in the 1960s as an outgrowth of the civil rights movement. They were a business-oriented response to the devastation of central-city neighborhoods caused first by urban renewal and then by the urban disorders of the 1960s. Philosophically, CDCs resonated with the kind of civil rights activism that was found in the cities of the northern United States rather than with the integrationist goals of the southern civil rights movement. Although southern civil rights leaders pressed for equal access and integration, northern activism focused more on the need for independent community restoration and self-help. CDCs were a mechanism for bringing about restoration through self-help.

A variety of forces were responsible for the destruction of many central-city neighborhoods during the 1950s and 1960s. Urban renewal, which began in the late 1940s and continued into the 1960s, was initially a major cause of the destruction. Urban renewal wiped out large numbers of housing units and businesses. Even though much of the housing torn down was substandard and in need of replacement, it had been home for many low-income residents, and its loss resulted in severe housing shortages in many communities. When coupled with racial discrimination, urban renewal left many minority households with few housing choices. Usually these choices were limited either to moving into newly built high-rise public housing developments or to pushing outward the boundaries of the black communities into previously all-white neighborhoods (Hirsch, 1983). The latter of these choices, which was often aided and abetted by unscrupulous realtors and lenders, resulted in white flight out of neighborhoods and subsequent redlining, which caused further neighborhood decay.

As central-city neighborhoods were losing their housing, they were also losing the businesses and industries vital to the local neighborhood economies. Urban renewal tended to put locally owned small businesses "out of business," and the subsequent neighborhood destruction and resulting decay caused owners of larger businesses and factories to rethink their loyalty to central-city sites. Responding both to declining neighborhood conditions and the lure of suburban opportunities, many owners folded their central-city operations and joined the migration of middle-class households to the suburbs.

The loss of jobs as businesses relocated to the suburbs resulted in what is known as a "spatial mismatch" between household locations and jobs (Kain, 1968), which further restricted employment options for the poor. Local business owners and others wishing to fill the gaps left by those who had fled were unable to do so because banks and other lenders redlined central-city neighborhoods, refusing to lend on mortgages, home improvement loans, or business loans. Thus, capital needed for both housing and economic development was either not available at all or in short supply. The combination of all the above forces produced neighborhood decline in city after city throughout the United States.

To many who were seeking to revitalize urban neighborhoods, advocacy organizing did not seem capable of addressing the issues associated with the steep economic declines that had occurred in many central-city neighborhoods. Attempts, even successful ones, to exert power and to gain control of neighborhoods seemed to be futile in places where resources had fled, disinvestment was rampant, and those responsible for decline were, often as not, located outside the neighborhood and were well insulated from advocacy organizing tactics. As Drier (1996) has recently noted, neighborhood organizations that have operated on their own have had limited successes because the resources needed to address the neighborhood's problems are not available at the local level and are often absent even at the city level.

Community revitalization of devastated neighborhoods, it was argued, calls for capital reinvestment in the community. If neither the government nor the private sector were willing to be engines of reinvestment, then, it was argued, the only way to bring about neighborhood revitalization was for the community itself to take on the tasks of reinvestment and redevelopment. Thus, self-help and community development, sometimes under the guise of "black capitalism" (see, e.g., Brazier, 1969), was born as a means of achieving neighborhood revitalization and community empowerment.

"CDCs can be seen as a response to the failure of the federal government under the Johnson Administration to institutionalize and fully support empowerment of the poor" (Keating, 1989, p. 8). Furthermore, CDCs were a response that differed from the radical activism, organizing, and advocacy that characterized the civil rights movement, particularly its northern component. CDCs, it was argued, were tangible ways of focusing activism and bringing about grassroots development through self-help efforts.

But they can also be thought of as a mechanism for blunting the more radical organizing alternatives. Proponents of CDCs argued that advocacy organizing led nowhere and that neighborhood development would only occur through the application of business principles and a stimulation of market forces in neighborhoods where the private sector had fled. In this light, CDCs can be thought of as a conservative, market-driven approach to neighborhood development.

The focus of early CDC efforts was on economic revitalization through job creation (Stoecker, 1997). The first CDCs numbered about 100, tended to be located in larger cities, and relied heavily on federal and foundation funding for program development. The Bedford-

Stuyvesant Restoration Corporation in the devastated Central Brooklyn area of New York is generally credited with being the first modern-day CDC (Bratt, 1989; Pierce & Steinbach, 1987).

By the 1970s, a second wave of CDCs was forming. This happened as a result of changing federal urban policy brought about by the Community Development Block Grant (CDBG) program that was initiated with the passage of the Housing and Community Development Act of 1974. The CDBG program shifted the focus of neighborhood problem solving from federal agencies to local municipalities, which in turn frequently sought not-for-profit community organizations to propose and carry out community development initiatives.

The newly formed CDCs tended to focus on housing development, which they perceived to be a major need in low-income communities (Vidal, 1992). Funds to purchase property and for rehabilitation and new construction initially came from federal subsidy programs, such as Section 8, and from private capital accessed through antiredlining campaigns made possible by the passage of the 1977 Community Reinvestment Act.

CDCs have generally gotten favorable reviews (see, e.g., Bratt & Keyes, 1997; Mayer & Blake, 1981; Pierce & Steinbach, 1987; National Congress for Community Economic Development [NCCED], 1989; Research and Policy Committee, 1995). A national survey of CDCs doing housing development, conducted in 1989 by the NCCED, found that CDCs were producing between 23,000 and 33,000 units annually throughout the United States (NCCED, 1989). In Chicago, the Chicago Rehab Network has estimated that 17 local CDCs produced 10,600 units between 1980 and 1993, an average of 44 to 45 units per CDC per year (Weiner, 1993).

Despite their successes, CDCs are still struggling to fulfill their promise. Although CDCs have had well-publicized successes,[12] many continue to have difficulty in carrying out their mission, and recently some CDCs, which had been thought to be successful operations, have gone out of business.[13] Even some highly regarded CDCs have been unable to bring about the hoped for revitalization in the neighborhoods in which they are located. In Chicago, for example, Bethel New Life, a nationally known and highly respected CDC, produced more than 700 units of new or rehabilitated housing during the 1980s. Nonetheless, the West Garfield community area, the focus of Bethel's operation, has continued to decline and, during the same period, experienced a net loss of 17% of its total housing (Weiner, 1993). Even though it is likely that the decline of West Garfield would have been even more severe had it not been for Bethel's presence, it is clear that Bethel has yet been unable, despite all of its efforts, to ignite the redevelopment of its west-side neighborhood.

CDCs also have had difficulty in providing housing for families who are truly poor. To keep housing costs as low as possible, CDCs must rely on government subsidies. In the 1990s, the major sources of subsidy have been low-income housing tax credits and HOME, a program created by the Affordable Housing Act of 1990. As Nelson (1994) has shown, the subsidies from both of these sources, which must then be used in conjunction with financing from local lenders and loan pools, result in housing with rents well above the level that is affordable to very low-income families. Instead of creating low-income housing, most CDCs have turned to providing affordable units to more moderate-income households. However, CDCs often have difficulty attracting moderate-income households to the rental units they produce because the neighborhoods in which they work often are plagued by serious social issues such as crime, drugs, and violence. This is especially true for CDCs operating in big cities.

CDCs also have discovered that it is easier and more exciting to develop housing than to manage it. Preserving housing through good management is essential if the housing is to be maintained and not allowed to deteriorate over time. CDCs often lack staff with management skills and are frequently unprepared and ill equipped to deal with the social and economic problems of their low-income tenants and the neighborhood problems that affect the ability of the CDC to keep good tenants in their buildings. A recent national study (Bratt, Vidal, Schwartz, Keyes, & Stockard, 1998; Schwartz, Bratt, Vidal, & Keyes, 1996) concludes that "nonprofit sponsors face serious challenges inherent in the management operations . . . that if unattended could result in the demise of the subsidized production system that has been built up over the past two decades" (Bratt et al., 1998, p. 47). In Chicago, increasing problems associated with CDCs and management have resulted in two initiatives led by the United Way/Crusade of Mercy that were intended to address some of the more serious management issues. The first, a "best practices study" of low-income property management

(United Way/Crusade of Mercy, 1996), identified exemplary ways to do low-income housing management. A second "collaborative initiative" (Brown, 1997) encouraged CDCs to work with local social service providers in ways that strengthen neighborhood capacity while providing much-needed services to tenants.

A stinging critique of the CDC movement has been made by Randy Stoecker[14] (1997), who argues that CDCs are not really bottom-up community-controlled organizations and that by focusing too narrowly on bricks and mortar, they have abandoned the social aspects of neighborhood development. Stoecker argues that for the most part, CDCs are too small, are grossly underfunded and thus too poor, and are staffed by inadequately trained and inexperienced individuals who, although well meaning, are incapable of successfully carrying out the tasks facing them. Noting that most of the people employed by CDCs are not community residents but instead are college-trained individuals from outside of the neighborhood, he questions whether CDCs are really community based.[15] Thus, although purporting to be grassroots community organizations, working to empower community residents, Stoecker argues that CDCs are little different from social service agencies that use professional outsiders to deliver services to communities.

Most important, Stoecker (1997) contends that by focusing on physical development, CDCs ignore the serious social and economic issues facing urban neighborhoods. Echoing arguments made by Piven and Cloward (1993) about antipoverty spending in general, he suggests that the focus given by government and business to the CDC approach to community revitalization may be an attempt to blunt more radical initiatives and that "government funding of CDCs may be most useful for maintaining social order" (Stoecker, 1997, p. 7).

Instead of CDCs as they are presently constituted, Stoecker (1997) proposes a separation between locally controlled community organizing and CDC-based community development. He would accomplish this by creating a small number of high-capacity, non-neighborhood-based multilocal CDCs in each city. These bigger CDCs would be capable of building and renovating large numbers of housing units. The separation of CDCs from their neighborhood base would have the effect of freeing advocacy organizations at the neighborhood level to pursue the goals of community empowerment without endangering a community's funding for housing and economic development.

As might be expected, Stoecker's (1997) views have not gone unchallenged by supporters of CDCs. Although agreeing that Stoecker has raised some important issues, both Bratt (1997) and Keating (1997) reject his conclusions that the CDC model is flawed. Bratt believes that CDCs have not been able to show their true potential because they have been hampered by scarce public resources, and Keating questions whether a broad-based social movement would actually occur if the responsibility for physical development was removed from neighborhoods and more community organizing was promoted.

Whether or not Stoecker (1997) is too critical of CDCs, he has nonetheless raised two important questions: What has been the real impact of CDC efforts, and what is or should be the relationship between community development and community organizing? In a review of community economic development activities, Giloth (1988) notes that there have been relatively few evaluations of CDCs and community economic development. Although he argues that "community based economic development promises to help poor communities" (Giloth, 1988, p. 349), he finds only limited examples of success. Likewise, Keating (1989) concludes that "without increased and sustained support, it is not clear that CDCs can really provide the housing, employment, and services necessary for the revitalization of urban neighborhoods" (p. 8).

Avis Vidal, who previously was director of the Community Development Research Center at the New School for Social Research and is now at the Urban Institute, has extensively studied community-based approaches to addressing urban problems and has written a "longer view" essay in the *Journal of the American Planning Association* (Vidal, 1997) about the future prospects for CDCs. In it, she identifies three issues facing CDCs. She says that CDCs and their supporters need to learn how to do the following:

1. effectively engage in an expanded range of activities,
2. adapt their structures to accommodate new rules and diminished resources for organizational support, and
3. articulate a new vision that can energize support for a diversified agenda (Vidal, 1997, p. 434).

All three of these are important but will be difficult to accomplish. Vidal feels, however, that accomplishing the third will be particularly difficult. According to her, diversification of the CDC agenda will involve the following:

1. broadening the mission of CDCs that are now primarily engaged in housing development to include nonhousing types of development that produce bankable economic assets for the community,
2. developing programs and activities that result in the creation of assets for residents, and
3. expanding into areas that involve the indirect provision of goods and services that link CDCs to other community institutions and that serve to broaden and deepen their community roots.

Diversification, as outlined by Vidal (1997), would make CDCs more like traditional community organizations. Many successful CDCs have always incorporated at least some of what Vidal suggests is needed. Many successful CDCs incorporate community organizing into their housing or economic development work, and others work closely with advocacy groups in their neighborhoods. More recently, CDC leaders have begun to recognize that the lack of social services and job training for their own residents and for others in their neighborhoods threatens the viability of their developments, and they have explored the possibility of expanding their efforts to include these activities. Yet, for the most part, CDCs have been hesitant to invest too much time, manpower, and money into efforts that they see as peripheral to their main mission—physical community development. But if Vidal is correct, then CDCs may be unable to have a meaningful impact in their communities unless they are willing and capable of broadening the scope of their involvement.

Vidal (1997) suggests that there are a variety of ways in which CDCs can expand the scope of their own engagement in the community and also create linkages with other community organizations. These include becoming involved in community policing programs, developing links with social service agencies operating in the community, participating in school reform activities, and undertaking, perhaps in partnership with other organizations, programs of job training and placement. To do these things, Vidal believes that CDCs will have to take on new roles, such as being brokers, negotiators, developers of shared agendas, and networkers, and they will have to strengthen their outreach efforts to the community.

Although continuing to recognize the central importance of CDCs as agents of community revitalization, a recent assessment of community development needs, performed by the Chicago United Way/Crusade of Mercy, asserts that if low-income neighborhoods are to be truly revitalized, then issues beyond those normally addressed by CDCs need to be looked after (United Way Needs Assessment Committee for Community Development, 1998). The writers of the assessment contend that "the best ways to build community are to increase civic involvement and to help young people reach their full potential as adults" (United Way Needs Assessment Committee for Community Development, 1998, p. 1). Although the writers go on to say that the best community-based organizations have always incorporated civic involvement and youth development into their work, they argue that there is a critical need to more fully incorporate these two items into the traditional housing, economic development, and community safety concerns of community groups in general and CDCs in particular.

*   *   *

A decade or so ago, it was thought that community development corporations and the work they did were sufficient mechanisms for revitalizing neighborhoods. Community organizing, particularly advocacy organizing, was thought to be counterproductive because organizers often targeted those groups with resources needed by the community. These groups, it was argued, were "turned off" by the strident, "nonbusiness" approach of the organizers and their community groups. Planning also was viewed as being not particularly relevant to community revitalization because the basic issues associated with improvement, better housing, more jobs, and safer streets were not only understood but were also areas that were only marginally addressed by the land use–oriented field of planning.

But today, the issues of community revitalization are no longer seen as so straightforward and are thought of as being quite complex and interrelated. Organizing has reappeared as a necessary element in a community development strategy, although it must work with and not against the activities of the CDCs. And planning is also no longer seen as irrelevant because there is an increasing need for the development of innovative strategies—short, mid, and long term—that address the complexity of neighborhood revitalization and serve to guide community-based activities.

Thus, community organizing, advocacy organizations, neighborhood associations, community development corporations, and community-based planning are all necessary components for the fashioning of mechanisms for community empowerment and neighborhood revitalization. Planners must learn to work with both community-based organizations as well as CDCs if they are interested in promoting notions of community empowerment and community-based revitalization.

The next chapter sets the stage for the four case study chapters that follow. In this chapter, I have attempted to lay out the context in which community and neighborhood planning and development occurs. It is the larger world of the city—the metropolitan area, the state, the nation, and the world—that works to shape what is possible in the neighborhood context. Failure to consider the broader context can lead to failure at the local level, and even well-thought-out and applied neighborhood strategies can fail in the face of external forces.

## NOTES

1. The section draws on material that originally appeared in Peterman (1996).

2. Conservatives also view community organization as the means for empowerment, but it is through organization that individual rights or power is obtained and maintained.

3. This is not to suggest that all empowered neighborhoods require the existence of formal not-for-profit community organizations. In many places, the mechanism of community control is informal, although just as strong. In others, community organizations are ephemeral, arising in times of crises and then lapsing as the crisis subsides.

4. For example, in 1987, Chicago's mayor, Harold Washington, established a "blue ribbon commission on public housing reform." Among the 25 or so members of this commission were 3 residents of public housing. Although the mayor died before the commission completed its work and thus the final report of the commission was never implemented, it was never clear from the outset what actual powers for making changes, if any, had been given to the commission. Whatever power the commission would have had if the mayor had lived, it was obvious that the public housing residents were vastly outnumbered on the committee, and their input did not count for much when it came to the commission's deliberations.

5. Residents at a public housing development in Toledo in the late 1970s had begun to organize to address inadequacies in the management of the housing by the Toledo Metropolitan Housing Authority. When they complained to the Toledo Metropolitan Housing Authority that their garbage was not being picked up often enough and that rats

were being attracted to overflowing garbage cans, the authority offered to present a series of seminars on how to properly pack a garbage can. Thus, the authority attempted to
avoid the real problem by manipulating the residents into thinking that they themselves
and their sloppy garbage can–packing practices were the problem.

6. When negotiating with the Chicago Housing Authority, the resident management corporation at Leclaire (see Chapter 7) attempted to gain maximum control by negotiating a formal contract with the Housing Authority. According to the terms of the
contract, the Housing Authority would provide funds to cover the monthly budget of the
development, and the resident corporation would submit monthly reports to the Housing Authority detailing how the moneys were spent. Subject to the relevant rules and
regulations, all decisions regarding the development, including employment, certifica-
tion of tenants, rules regarding tenant responsibilities, external contracts, and security,
would be the responsibility of the resident corporation. The Chicago Housing Authority
was never comfortable with yielding this much control to the residents and eventually
"recaptured" the development and voided the contract.

7. An example of the kind of confrontational activism associated with Alinsky-style
organizing occurred in the 1980s in Chicago when the Chicago Bears football team proposed building a new stadium on the near west side of the city. At the time, the Interfaith
Organizing Project organization was actively trying to preserve the neighborhood from
redevelopment. Opposed to the new stadium, they organized a demonstration at the upscale suburban home of the Bears' owner. Some of the demonstrators played touch football in front of the owner's home. Others went house to house asking if the residents
knew of any low-cost housing in the area because they were about to be displaced by the
actions of their neighbor. This demonstration received considerable television coverage
and was one of many actions that subsequently resulted in the Bears withdrawing their
proposal for a stadium.

8. After an absence of many years, the Industrial Areas Foundation (IAF) returned
to Chicago in the mid-1990s at the request of Chicago's Roman Catholic Archbishop
Bernadin. With support from the archdiocese and other religious bodies, the IAF has
mounted an organizing campaign intended to unite the various communities of the city
and its suburbs into a potent citizen force. As of this writing, the organizing effort has succeeded in bringing together some 10,000 "Chicagoans" in a mass show of power but has
yet to define an action agenda.

9. A thorough discussion of the issue of redlining and the use of Home Mortgage
Disclosure Act and the Community Reinvestment Act as reinvestment tools is covered by
Squires (1994).

10. Kotler (1969) argued that modern-day cities had grown by engulfing small, self-
governing towns. For example, he noted that Roxbury, Massachusetts, had been settled
as a town in 1630 and had remained a self-governing unit until 1868, when it was annexed
by Boston.

11. In 1979, the National Association of Neighborhoods adopted a neighborhood
platform proposing that neighborhood residents be empowered to establish representative neighborhood governments that would have at least the following powers: the ability to raise tax revenues; incur bond indebtedness; enter into interjurisdictional
agreements; settle neighborhood disputes; contract with the city or with private providers of services; conduct elections; sue and be sued; determine planning, zoning, and land
use; exercise limited eminent domain; undertake public investment; provide public and
social services; and operate proprietary enterprises (National Association of Neighborhoods, 1979).

12. For example, the Banana Kelly Community Improvement Association, started in 1977 as a struggling urban homesteading effort, is now a $7 million, 100-employee community development corporation (CDC) that manages 26 buildings with more than 1,000 tenants.

13. The Chicago CDC community was particularly surprised when in 1995, one of its better-known members, Peoples Housing, went from a seemingly viable organization to bankruptcy within a few short months.

14. Stoecker (1997) claims that he is not an outsider critic of the CDC movement and contends that he is a friend of community development having "for over a decade . . . followed and worked with community development corporations" (p. 1).

15. Most CDCs have boards of directors that are made up of a majority of community residents. Stoecker (1997) contends that these boards usually exert only minimal influence on the operation of the CDCs and are easily manipulated by the director of the CDC and his or her staff.

# 4

# CHANGING URBAN STRUCTURE

*Implications for Neighborhood Development*

Many prescriptions for neighborhood revitalization are based on an assertion that neighborhoods have within themselves the capacity to determine and bring about their own futures. Revitalization occurs, it is argued, when neighborhoods recognize and use this capacity for positive change.

This notion of neighborhood self-determination assumes that external forces that affect a neighborhood can be overcome through the pooling of the neighborhood's resources. Such an assumption ignores the structural relationships of power in urban areas and dismisses as unimportant the reasons why a neighborhood needs to be revitalized in the first place. It further ignores the structural changes that, since the beginning of the 1970s, have weakened the power of central cities, have changed the ways in which revitalization can be financed, and have made nearly all the issues associated with neighborhood revitalization more complex than in the past. Although self-help efforts are important and probably a necessary component of neighborhood development, they are unlikely to succeed without taking into account the context in which neighborhood development must currently occur.

As cities have become more decentralized, as the balance of political and economic power has shifted away from central cities and toward the suburbs, and as poverty has become increasingly concentrated in the core of our urban areas, it has become increasingly difficult for cities to address the needs of their citizens and the places where they live. Efforts to increase the urban tax base and to decrease the costs of providing social services usually have meant that city governments, supported by real estate interests, have focused on downtown development and gentrifying areas of the city in the hope of attracting or retaining the middle class rather than addressing the needs of the poor and of distressed neighborhoods. Given these trends and the current condition of our cities, it might even be argued that any attempts to bring about urban revitalization using an approach that focuses on neighborhood development and relies on grassroots community-based development efforts are misplaced.

Before turning our attention to the several grassroots revitalization efforts that make up the core of this book (Chapters 5-8), we therefore need to first discuss, even if briefly, some of the specific forces that negatively affect neighborhood redevelopment efforts. Three of these—the locus and control of political and economic power in cities, growth coalitions and their impacts on urban development, and the changing spatial structure of urban areas—are discussed in the following paragraphs. These are not separate issues and are interrelated to a considerable degree. Although proponents of neighborhood redevelopment frequently claim that a neighborhood or community can control its own destiny, forces such as these three often work against neighborhood control. Thus, some understanding of them ought to be helpful in our exploration of the limits and opportunities for neighborhood redevelopment. Minimally, at least, the external forces shape the playing field on which redevelopment efforts occur.

## POWER, ITS ROLE IN CITIES, AND ITS RELATIONSHIP TO NEIGHBORHOODS

This book is an attempt to make a realistic appraisal of the potential and limits of grassroots organizing, planning, and development. Doing this requires identifying the role that neighborhoods play in the complex web of political, economic, social, and cultural relationships that come together in a city. It is not possible, for example, to talk about the revitalization of an African American neighborhood without addressing the issue of race and all its political, economic, social, and cultural implications. Nor can we discuss the revitalization of older working-class white neighborhoods without addressing the loss of employment opportunities that are the result of forces that shape the job market for the neighborhood's worker population but are quite external to the neighborhood.

Community activists and organizers have long argued that by coming together and demanding their "rights," neighborhood people constitute a powerful force for the improvement of their neighborhood (see, e.g., Speeter, 1978). This notion of neighborhood power, however, has been challenged by two Chicagoans—Charles Bowden and Lew Kreinberg—in their book *Street Signs Chicago: Neighborhood and Other Illusions of Big-City Life* (Bowden & Kreinberg, 1981). Bowden and Kreinberg contend that neighborhoods are not what makes cities work and indeed are not even central to the processes that shape urban life. Although they agree with others that power is the key to getting things done, they state that power is not often found in neighborhoods, at least not in the neighborhoods of Chicago.

> Peel back the labels of the city, peel back community, urban renewal, peel back neighborhood, and then you will find the power to make and break buildings, lakes, rivers, and people.
>     Power moves in currents above and beyond places called neighborhood; power moves in money sliding down LaSalle Street to the Chicago Board of Trade, cruises around in limousines waving real estate plans, flashes by in electric lines nourished by nuclear fires, explodes in metal cylinders that drive automobiles, trucks, boats. (Bowden & Kreinberg, 1981, p. 40)

Both Bowden and Kreinberg are trained historians, longtime observers of Chicago, and Kreinberg has spent most of his adult life toiling as a community organizer in a variety of Chicago neighborhoods. Thus, when these two assert that "neighborhoods are not the tail that wags the dog. They are just located at the ass end" (Bowden & Kreinberg, 1981, p. 32), we can smile at their colorful Chicago-style analogy, but we should take them seriously. Too many people, Bowden and Kreinberg argue, are fooled by politicians who claim that neighborhoods are the fundamental units of a city. To Bowden and Kreinberg, this is nothing more than a smoke screen intended to hide the real meaning of cities and the real locus of urban power.

We are a nation of individualists who like to believe that we can control our own destiny. We are also a nation of organizations or associations, a characteristic of our country noted as early as the 1830s by the French observer de Tocqueville when he visited the then young United States.[1] The mix of rugged individualism, combined with our urge to form into associations of like-minded individuals, leads us to believe that what we strive for as individuals is even more achievable through group association. We look, therefore, to accomplish private goals through group action.

We are uncomfortable with any suggestion that we cannot achieve our goals or that our local associations cannot help us. We choose to ignore the obvious—that key decisions affecting our lives and, by extension, the lives of the neighborhoods in which we live frequently are beyond our control—and many decisions about us and the state of our neighborhoods are made by people who are not our neighbors.

Bowden and Kreinberg (1981) are skeptical about the possibilities of grassroots political power. They argue that

> Chicago, for example, describes itself as a city of neighborhoods. This is not true.
>
> Chicago is a city of factories, railroads converging on the bog, bankers counting the take, steel, gadgets, profit and loss. Neighborhoods come so far down this list that they are off the page. Nobody came to Chicago to found a neighborhood or to save the lake shore for swell parks. (Bowden & Kreinberg, 1981, p. 32)

## GROWTH COALITIONS, CITIES, SUBURBS, AND URBAN NEIGHBORHOODS

In many ways, Chicago, as is the case for many cities in the Northeast and Midwest, reached its peak as an industrial engine at the end of the 19th century and beginning of the 20th century. It has been in decline ever since. By the time Carl Sandburg (1916/1994) labeled Chicago as the "hog butcher of the world," industries had already begun moving to what we now call "greenfield sites" at the edge of the city or at least outside of the built-up areas of the city. Cyrus McCormick's choice, following the Great Fire of 1871, to relocate his farm implement factory to a site near what is now the Cook County Jail on the west side of Chicago and George Pullman's decision to locate his palace (sleeping) car factory on rural land south of the city and to build a model worker's town next to it on the shore of Lake Calumet were typical of industrial development activity at the end of the century and are consistent with a pattern of central-city abandonment that continues to this day.[2]

Early on, new industrial development at the edge of the city, such as the location of Chicago's Union Stock Yards on the city's south side, mattered little because the city merely expanded outward. These newly developing areas were annexed and their tax bases captured by the city. But by the turn of the century, the incorporation of formal suburban communities had begun to stop or at least severely limit the ability of cities to expand outward. Thus, as industry moved further and further away from the center of the city, its new location was more and more likely to be a suburb.

During the Great Depression, there was little building anywhere. As boom times returned with World War II and continued into the postwar era, it was the suburbs, not the cities, that benefited. Squires, Bennett, McCourt, and Nyden (1987), for example, point out that by 1947, Chicago's manufacturing base was moving to the suburbs, and its economy was beginning to shift away from the manufacturing sector and toward the service sector. At the same time, housing construction was booming in the suburbs as returning World War II servicemen married, started families, and left their old neighborhoods for a new suburban house with a picket fence and a new suburban job.

As early as the 1930s, civic and downtown leaders in Chicago worried about the future of the city and about their investments in the city's "Loop" (see Hirsch, 1983). The massive urban renewal and highway building projects of the late 1940s and 1950s were efforts to revitalize the central core of the city. These efforts were encouraged through the formation of what Logan and Molotch (1987) call "growth machines," an "apparatus of interlocking pro-growth associations and governmental units" (p. 32). Growth machines, driven by landowners who want to protect their investments and local government officials who want to protect the tax base of the community, are the real power elite in most if not all urban areas.

Members of the growth machines of the postwar period cared little for neighborhoods and viewed the run-down neighborhoods that ringed downtown as a threat to a city's progress. In Chicago, Mayor Richard J. Daley embraced the growth machine concept by undertaking massive public works projects such as expressway building, the construction of O'Hare Airport, and the construction of the McCormick Place convention center along the city's lakefront. All this focus on growth and big-ticket development projects did little to assist the neighborhoods, which, for the most part, continued to decline as they had been doing since the Great Depression.

Suburban areas have their own local growth machines as well. During the past 20 to 30 years, satellite urban centers or "edge cities" (Garreau, 1991), fostered by pro-growth coalitions, have drained away much of the retail and office strength of the central city. Today, in "Chicagoland," people are just as likely to shop and work in suburban Oak Brook or Schaumburg as they are in the Loop. A city government facing this reality and obsessed with bringing people back downtown and stopping the decline of its central core is likely to have little energy for addressing the decline of its neighborhoods.

Today, cities such as Chicago are not the powerful entities that they once were. Strapped for resources, cities do their best to compete with both their own suburbs and with other cities in their region. Even pro-neighborhood mayors, such as Chicago's first black mayor, Harold Washington (1983-1987), can feel compelled to take actions that turn out to be detrimental to neighborhoods. In Washington's case, he supported the construction of a new stadium for the Chicago White Sox, fearing that if he did not, the team would be lost to a Florida city. The eventual construction of the stadium, which will be discussed in Chapter 7, resulted in the destruction of a stable neighborhood of working-class African Americans and left a public housing neighborhood in complete isolation.

As we consider what cities can do to assist neighborhood development, we should again reflect on what Bowden and Kreinberg (1981) have to say:

Most efforts to fix the city (by neighborhood activists) hearken back to a time when . . . [the city] was said to work. . . . None of . . . [the] plans recognize that the rich resource base that built Chicago is diminished and that the price of doing business in the city has gone up. Nor do they recognize that even when things were cheaper Chicago did not do well by its citizens. (p. 19)

## THE CHANGING SPATIAL STRUCTURE OF CITIES: ISSUES OF CLASS AND RACE

But what are the implications of power, growth machines, and the changing location of jobs and commerce for cities and especially for their neighborhoods? Consider what it has meant in Chicago. The changes in metropolitan Chicago have been dramatic, leading to a new metropolitan form characterized by low-density sprawl and the growth of outlying edge cities. Although the outer suburban areas are congested and booming, the central city and close-in suburbs are emptying out. Old factories stand abandoned, and many neighborhoods have as many vacant lots as they have houses. The quality of life in many Chicago neighborhoods, particularly neighborhoods that are home to low-income households, has declined.

Consider the viability of neighborhoods as economic units. Nearly everyone is aware that during the past few decades, industrial employment in the Chicago region has declined significantly. But what is less well known is that nearly all of this decline has occurred within the city of Chicago itself, whereas the number of industrial jobs in the suburbs has declined only slightly.

In 1972, there were just over 1 million industrial employees in the Chicago metropolitan region (including northeastern Indiana). Of these, 430,100 were employees who worked in the city of Chicago. The city thus accounted for 43% of all industrial jobs in the region. By 1987, industrial employment in the region had dropped to just over 755,000, a substantial loss of almost 250,000 jobs. All but 40,000 of this net loss of jobs was in Chicago. Although industrial employment declined by 49% in Chicago, in the remainder of the metropolitan region, the decline was a mere 7%. Thus, industrial decline was a phenomenon singular to Chicago. In the suburbs, it was more or less business as usual.

Did service-sector jobs act to offset the decline in industrial-sector jobs, as is generally believed? The growth in service jobs in the Chicago region was indeed remarkable, going from 257,000 jobs in 1972 to 599,000 jobs in 1987, an increase of 342,000 jobs. This means that considering employment in both the industrial and service sectors, there was a net increase of about 90,000 jobs. But Chicago's share of the new service-sector jobs was small. Its service job base increased from 157,000 jobs in 1972 to only 242,000 jobs in 1987. Thus, the net change in Chicago itself was a loss of about 120,000 jobs.

Since 1970, Chicago has become less and less of a workplace, and this has affected the city's neighborhoods in several critical ways. Closed factories mean vacant and abandoned buildings, which become community eyesores and potential sites for criminal and other antisocial behavior, particularly in low-income neighborhoods. Loss of community employers can also have psychological effects on a community, making it seem like a "bad" place to live. Often these effects are real because the local businesses, organizations, and agencies can no longer rely on major employers for support. The families and organizations in the community also can be affected because as local jobs are lost, workers must make time-consuming commutes to suburban places of employment and thus have less time to be with their families and to participate in community activities.[3]

Shopping opportunities have also declined, even to the point of disappearing in some neighborhoods. In 1972, Chicagoans made up 42% of the metropolitan population, and Chicago's retail establishments accounted for 40% of the metropolitan sales. Sales and population were nearly in balance. By 1987, Chicago's share of the population had declined to 36%, but its share of the retail sales had declined to only 24% as more and more city shoppers were lured to suburban sites. Although the volume of sales in dollars in Chicago increased by 60% during that period, the suburban increase was 234%. These figures are even more dramatic considering that there has been considerable growth in retail activity in the central core of Chicago, especially in the North Michigan Avenue and River North areas of the core. Sales volume in the neighborhoods has remained nearly constant in actual dollar amounts, and in many neighborhoods, sales have declined.

These economic figures indicate an overall worsening of the quality of life in Chicago's neighborhoods, but they are not the only trends that could have been highlighted. The past 20 years have seen a similar decline in many of Chicago's neighborhood institutions, both private and public, and the amount and quality of resources available to neighborhoods, especially education, parks and recreation, and a host of social services. Coupling these losses with the decline of many central-city job opportunities, it has become increasingly difficult for families to live and prosper in the city, and many who can afford to have chosen to leave for better opportunities in the suburbs. The net result in many neighborhoods, as middle-income families and even the working poor leave, is the concentration of families and individuals with serious social and economic problems who are trapped in increasingly hostile environments.[4]

The restructuring of the economy has resulted in worsening social and economic conditions in neighborhoods throughout the United States. Cuts in state and federal assistance, particularly severe since the onset of the Reagan years in the 1980s, have exacerbated the problem, leaving many central-city neighborhoods devastated. The results have been an escalation of gang activities, crime, drug dealing and use, and domestic and random violence. It is still too early to gauge the impacts of "welfare reform" that has swept the country, but in low-income neighborhoods, it is likely to exacerbate an already serious set of problems.

It is in these highly stressed neighborhoods that community organizations and community development corporations struggle to improve the quality of life for low-income and working-class households. These organizations are almost always understaffed, underfunded, and forced to pay low salaries to young, often inexperienced, though committed individuals. Essentially, they have too few of the resources needed to carry out their missions. But in many neighborhoods, these are the only active organizations. It is a big task they shoulder, and it should not be surprising that their successes are few despite their considerable efforts to produce results.

## RACE AND CITIES

Added to the above trends that are making it increasingly difficult to revitalize urban neighborhoods, we must also consider the serious inequalities relating to race. Although many middle-income minority households have joined their white counterparts in moving to the sub-urbs,[5] poor minority households, especially blacks and Latinos, have continued to suffer the effects of discrimination and neighborhood dis-investment. They remain locked in areas of increasing poverty and decreasing opportunities. Unlike some of their white counterparts, such as those in Boston and chronicled by Gans (1962), there is little hope that the children of these minority poor will be able to move to a better neighborhood and to improve socially and economically. They seem doomed to remain as part of Wilson's (1987) "underclass."

In Chicago, although there was a loss of nearly 800,000 residents between 1950 and 1990, the African American population more than doubled from approximately 490,000 in 1950 to almost 1,000,000 in 1990. In 1950, African Americans made up only 14% of the Chicago population. In 1990, they were 39% of the total. Historically, Latino households were only a small but significant portion of Chicago's eth-nic mix, but by 1990, they accounted for 29% of the population. Thus Chicago, like many cities, is a place where minorities have become the majority.

The concentration of poor minority populations has had a devastat-ing effect on the images of cities and on their efforts to attract middle-income families, and this effect extends well beyond the underclass neighborhoods. Suburbanites are often reluctant to even visit the city, believing it to be a dangerous place filled with people unlike them. Although the negative image of cities is a reflection of prejudices of those people who perpetuate it, it is nonetheless a problem and a bar-rier to be faced when addressing neighborhood revitalization. And it becomes more than just a problem when prejudice is translated into actions such as mortgage redlining and racial steering in the sale or rental of housing (Squires, 1994).

The continuing existence of a discriminatory dual housing market that treats white and minority home seekers differently (Bradford, 1979; Squires et al., 1987; Turner, Struyk, & Yinger, 1991) despite 30 years of experience with the 1968 Federal Fair Housing Act has greatly exacerbated efforts to sustain viable urban neighborhoods. Its exis-tence has helped to ensure the continuing concentration of minorities in the central city and their relative absence in the suburbs. Mortgage lending discrimination (Dedman, 1988; Squires & Velez, 1987) and insurance redlining (Squires, Velez, & Taeuber, 1991), which accompany the dual housing markets, are also critical barriers to the viability of minority neighborhoods and must always be considered in any revitalization strategy. Reinvestment in minority neighborhoods has been slow in coming, despite the existence of federal laws such as the Fair Housing and Community Reinvestment Acts, which appear to mandate equal treatment of minorities and the neighborhoods in which they live. Unfortunately, enforcement of both these laws by the federal government has been lax and, for the most part, left to not-for-profit and community organizations to initiate.

*   *   *

Given the above discussion, it would seem that doing neighborhood development and revitalization involves much more than calling the neighborhood folks together, forming a local grassroots organization, developing a plan, and implementing it. Just what and how much more is needed is at least part of what this book is about. Neighborhood revitalization may indeed be possible, but the task is often far more difficult than is imagined.

We are now, however, ready to proceed to the case studies of neighborhood planning and development that appear in Chapters 5 through 8. They are narratives of situations that I have experienced during my career as the director of a neighborhood assistance center, first at the University of Illinois at Chicago and more recently at Chicago State University. Thus, they all are Chicago examples. In two of the case studies, I played a key role in the planning and development process. In the other two, I was one of many individuals involved, and my role was not central to what happened. Three of the cases involve community organizations, the fourth a community development corporation. However, the distinction between types of groups is not always clear. Some of the community organizations were involved with development, but a major focus of the community development corporation is on community organization. As will be seen, the cases do not lay out a set formula for neighborhood revitalization. Nor do they suggest that any one type of community organization is ideal. Rather, they show that there are many routes to neighborhood improvement.

The cases are presented in ascending levels of success. That is, the hoped for outcome of the planning process described in the first case situation did not materialize. The following cases, however, increasingly become stories of success. It is my intent to use these studies to demonstrate not only how neighborhood planning and development are carried out but also how successful outcomes can be achieved.

## NOTES

1. Alexis de Tocqueville (1835/1945) was primarily concerned with what he called political associations, but he noted that "in the United States associations are established to promote the public safety, commerce, industry, morality, and religion" and that "there is no end which the human will despairs of attaining through the combined power of individuals united into a society" (p. 199).

2. The industrial development of nearly all of northwest Indiana grew up on "greenfield" sites. Eldridge Gary, like George Pullman, for example, chose to locate his steel mill (U.S. Steel) on what was then sand dunes and to build a town around it.

3. Data from the 1990 Census of Population and Housing show that workers in the neighborhoods of Chicago have the longest average commuting times of any workers in the region. For example, workers residing in the south-side community area of Chatham travel an average of 37.8 minutes to work, whereas workers in suburban DuPage County travel an average of only 25.2 minutes. The suburbanization of work sites has rendered obsolete the conventional wisdom that suburban workers make long commutes to jobs in the central city.

4. Goldsmith and Blakely (1992, p. 46) report that between 1970 and 1980, the poverty population in the nation's 50 largest cities grew by nearly 12%, whereas the overall population of these cities declined by 5%. Furthermore, these poor have increasingly become concentrated in high-poverty neighborhoods. Cities with high poverty rates also are likely to contain a significant number of high-poverty census tracts (Goldsmith & Blakely, 1992).

5. A recent study of housing patterns in metropolitan Chicago (see Leachman, Nyden, Peterman, & Coleman, 1998) has shown that although the number of black and Latino households in Chicago's suburbs is increasing, the pattern of suburbanization is such that segregation is being reinforced. That is, blacks and Latinos are moving to suburbs that already contain significant numbers of minorities and do not seem to be making moves that are consistent with their economic status. Furthermore, suburbs with high percentages of either black or Latino households tend to be places where the number of jobs has been declining and where the tax base is low. Thus, there is evidence that the patterns of decline that followed white flight and resegregation of urban neighborhoods are continuing in the suburbs.

# 5

# A COMMUNITY
# CONFRONTS GENTRIFICATION

In the past 20 to 25 years, gentrification, the upgrading of housing and neighborhoods in central cities by real estate speculators and urban pioneers, resulting in the displacement of the neighborhoods' former and usually lower-income residents, has occurred to some degree in nearly every U.S. city. Civic leaders, especially politicians and real estate magnates who make up the main components of urban growth machines (see Logan & Molotch, 1987), favor and promote gentrification because they feel it is necessary to the overall financial well-being of cities and because they almost always personally benefit from it. Community people and their organizations generally oppose gentrification because it brings higher rents and taxes that result in many longtime residents being squeezed out of the neighborhood.

AUTHOR'S NOTE: This chapter draws on material that originally appeared in Peterman and Hannon (1986).

In many ways, gentrification can be thought of as the continuation of efforts to revitalize central cities that began with urban renewal in the late 1940s and 1950s.[1] However, as a process, gentrification appears to be less heavy-handed than urban renewal, possibly because it is the result of many individual private decisions rather than a few public policy decisions. In addition, it is more unpredictable, meaning that one cannot necessarily determine in advance whether a neighborhood will gentrify. Nonetheless, once gentrification begins in a neighborhood, the impacts are highly significant (see, e.g., DeGiovanni, 1983).

This chapter is about the gentrification of a relatively small area on Chicago's north side, West DePaul. The case study focuses on a community organization in the area, Concerned Allied Neighbors (CAN), and its attempts to respond to neighborhood changes resulting from gentrification. CAN initially saw gentrification as a threat and wanted to do something that would limit the negative impacts to existing renters and homeowners. So it set out to develop a plan or set of strategies that would help to preserve as much of the old neighborhood as possible. But its efforts failed when CAN itself became a victim of gentrification.

## WEST DEPAUL AND CONCERNED ALLIED NEIGHBORS

The West DePaul neighborhood lies just east of what has become Chicago's vibrant, upscale, Clyborn corridor, consisting of trendy stores, restaurants, and lofts (see Figure 5.1). The revitalization of the corridor is one of Chicago's "success" stories of the 1980s. Once a drab and declining industrial zone, the area is now alive both day and night.

Like the corridor, West DePaul also has changed. In 1980, it could best be described as a working-class neighborhood consisting mostly of small single-family homes and a few small apartment buildings (see Figure 5.2 and Table 5.1). Throughout the approximately 50-block neighborhood, there were small factories, many interspersed on blocks with housing. A spur railroad track ran along the eastern edge of the neighborhood that was used, albeit infrequently, to deliver shipments to factories in the neighborhood and to a few other factories further to the north.

Housing quality and the income of households declined from east to west. The median family income varied from $19,565 in the eastern-

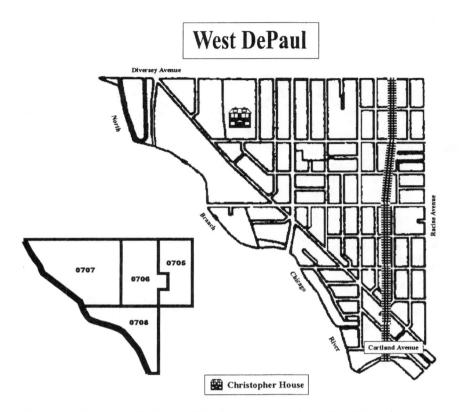

**West DePaul**

Diversey Avenue

North

Branch

Chicago

River

Racine Avenue

Cortland Avenue

0707    0705

0706

0708

🏠 Christopher House

**Figure 5.1.** Map of the West DePaul Neighborhood Boundaries: North Branch Chicago River, Diversey, Racine, and Cortland Landmarks (Christopher House and Milwaukee Road Spur)

most census tract to only $8,857 in the western-most tract. The median family income, however, in all three census tracks was below that of the city of Chicago, which in 1980 was $28,775. A settlement house, Christopher House, was centrally located in the neighborhood and offered a variety of social services and program opportunities for neighborhood residents.

According to official city designation, West DePaul is part of the Lincoln Park community area,[2] but in 1980 it bore little resemblance to the rest of Lincoln Park. Beginning in the early 1960s, starting first in the "Old Town" area,[3] Lincoln Park had been the first of Chicago's older

**TABLE 5.1** Demographics of the West DePaul Neighborhood

|  | Census Tract | | | |
|---|---|---|---|---|
|  | *705* | *706* | *707* | *708* |
| **1980** | | | | |
| Total population | 2,148 | 2,097 | 2,927 | 1,136 |
| % black | 2 | 1 | 10 | 6 |
| % of Spanish origin | 16 | 38 | 32 | 35 |
| % 13 years old and younger | 14 | 21 | 26 | 19 |
| % 65 years old and older | 12 | 9 | 14 | 9 |
| Median family income, 1979 | $19,565 | $17,538 | $8,857 | $14,648 |
| % income below poverty level | 12 | 13 | 34 | 16 |
| % white-collar workers | 59 | 47 | 37 | 32 |
| Population per household | 2.1 | 2.6 | 2.5 | 2.7 |
|  | | | | |
| Total housing units | 1,169 | 948 | 1318 | 437 |
| % condominiums | 2 | 0 | 0 | 0 |
| % built 1970 or later | 0.048 | 1 | 3 | 6 |
| % owner occupied | 0.244 | 22 | 16 | 23 |
| Median value: Owner units | $53,900 | $45,000 | $31,900 | $32,500 |
| Median rent: Rental units | $199 | $165 | $100 | $158 |
|  | | | | |
| **1990** | | | | |
| Total population | 2,175 | 2,128 | 2,655 | 1,228 |
| % black | 2 | 1 | 24 | 13 |
| % of Spanish origin | 7 | 27 | 38 | 18 |
| % 13 years old and younger | 7 | 14 | 21 | 15 |
| % 65 years old and older | 14 | 7 | 12 | 15 |
| Median family income, 1989 | $71,596 | $47,188 | $12,909 | $28,417 |
| % income below poverty level | 0 | 14 | 37 | 28 |
| % white-collar workers | 90 | 66 | 52 | 70 |
| Population per household | 1.8 | 2.4 | 2.2 | 2.1 |
|  | | | | |
| Total housing units | 1,274 | 1,115 | 1,370 | 669 |
| % condominiums | 5 | 11 | 0 | 7 |
| % built 1980 or later | 5 | 30 | 0 | 24 |
| % owner occupied | 30 | 49 | 15 | 23 |
| Median value: Owner units | $247,200 | $289,000 | $125,000 | $413,900 |
| Median rent: Rental units | $695 | $536 | $228 | $441 |

NOTE: Data in this table are from the U.S. census as organized and reported in the *Local Community Fact Book, Chicago Metropolitan Area 1990* (Chicago Fact Book Consortium, 1995).

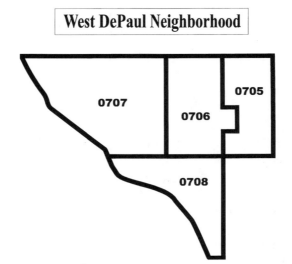

**Figure 5.2.** Census Tract Map of the West DePaul Neighborhood Tracts 705, 706, 707, and 708

residential areas to gentrify. The gentrification of Old Town expanded northward into "New Town" during the 1970s and had reached DePaul University's north-side campus area by 1980. Even though the area immediately adjacent to the university had gentrified, there was an abrupt change just west of the campus. In 1980, West DePaul looked very much like an old, nearly worn-out community that was deteriorating and would probably continue to decline.

West DePaul was so different from the rest of Lincoln Park that its primary community organization, CAN, had affiliated with the Lakeview Citizens Council (LVCC), an umbrella organization representing groups in the more working-class community area of Lakeview to the north.[4] Formed in 1975 from the coming together of several block clubs, CAN was a small but active organization made up of a mixture of homeowners and renters.[5] It was supported by Christopher House, which provided the salary for CAN's single staff person, a community organizer. Despite its smallness, CAN had been somewhat effective in halting neighborhood decline through its advocacy for park and recreation improvements, anti-gang programs, monitoring of city housing court activities, and support of the building of some successful subsidized housing units.

## COMMUNITY-BASED PLANNING IN A GENTRIFYING NEIGHBORHOOD

In 1981, the CAN community organizer began to notice that gentrification in Lincoln Park was edging westward toward West DePaul and that new development was beginning to occur at the southern end of Clyborn. Sensing that within a few years gentrification could become an issue in West DePaul, he tried to alert CAN leaders about what might be in store. He contacted the Voorhees Center for Neighborhood and Community Improvement, which I directed at the University of Illinois at Chicago, and I agreed to meet with him and several of the organization's leaders. The meeting was held in the back room of a neighborhood bar.

I made a brief presentation about the importance of community planning and the need to develop strategies for the future. The residents in the room seemed unconvinced that gentrification was a threat and argued instead that a far more serious problem was the ongoing deterioration of the neighborhood and local gangs. No action was taken during or after the meeting. I heard only once more from the organizer, at which time he told me that he had been unable to convince the organization that gentrification could become a problem and that long-range planning should be done in anticipation of neighborhood change.

I next heard from CAN in the spring of 1985. By this time, rehabilitation of homes and the displacement of long-term residents were taking place in the blocks just to the east of CAN's boundary. Several of CAN's leaders were now becoming concerned and were saying that they believed gentrification would begin in their neighborhood sometime in the next 2 to 3 years. Because of their growing concern, they had instructed their new organizer to seek help. Based on information he had picked up from the previous organizer, he contacted the Voorhees Center. After another initial meeting, we began almost at once to assist CAN in developing what was to be a plan or a set of strategies that would attempt to strike a balance within the neighborhood. By promoting the maintenance of a mixed socioeconomic population and providing affordable housing options, it was hoped that it would be possible to capture some of the benefits of gentrification while preserving the existing neighborhood.

Even as we began the planning process, it was apparent that the resident leaders of CAN were not in full agreement as to their goals for the neighborhood. Homeowners in the organization tended to be somewhat ambivalent about gentrification, seeing positive as well as negative benefits associated with it. Renters, however, were more certain about their opposition to gentrification, stating that they would be quickly displaced once gentrification started.

The planning process took about 8 months to complete and included physical surveys of existing buildings, interviews with community leaders, strategy sessions with CAN members, and several community meetings. By the time we were ready to formulate the components of a plan in February 1986, there were at least five proposals for new developments or redevelopments in the neighborhood. Gentrification, which the CAN leaders had thought was 2 to 3 years away, had arrived.

It can easily be argued that we were invited back the second time much too late for our planning assistance to be of much use in halting or even redirecting gentrification. The situation is, however, typical. Community-based organizations usually have very limited resources. The organizations must use these limited resources to address current problems facing the community. Thus, they are often unable or unwilling to engage in any mid- or long-term planning efforts.

Given the rapid changes that were already beginning to occur in the neighborhood, we recognized that any kind of conventional plan that might be developed would probably be obsolete by the time it was completed. Therefore, we decided to concentrate our efforts on identifying ways in which CAN could become involved in the process of neighborhood change and thus, it was hoped, be a force in helping to direct how change would actually occur.

Working with leaders of CAN, we set out to create a plan for community development that would include the organization and would rely heavily on input from CAN members and from existing neighborhood residents. We worked in a "participatory" fashion, meaning that whenever possible, work tasks were shared with CAN members. We gathered information about the neighborhood, surveyed housing quality, reviewed existing planning documents relevant to West DePaul, inter-viewed "key" community residents, and compiled a list of identified community needs. We met at least once every other week with CAN leaders and held three meetings at a local church that were open to all neighborhood residents. These meetings were used to identify existing community trends, set community goals and objectives for the immedi-ate and long-term future, and identify possible strategies for reaching the goals and objectives.

Initially, we felt that the planning process was proceeding smoothly. However, this feeling would not last long. We soon began to sense that problems loomed on the horizon. CAN members, in general, and the members of its planning committee, with whom we were working directly, began to express doubts about their ability to implement any of the community development strategies we had begun to suggest. Furthermore, we were even beginning to feel uneasy about CAN's commitment to the planning process.

Although CAN's leaders would verbally express their commitment to confronting the issue of gentrification, we found them to be more or less ambivalent about the changes that were going on in the neighbor-hood. Those who were homeowners felt that they would benefit from neighborhood improvements and rising property values. Both owners and renters felt that neither they nor CAN could really do much to affect existing neighborhood trends. Even though as an organization they were supporting a process whose goals were to promote neighbor-hood stability and preserve affordable housing, individually they seemed unable to become as involved as they had previously when the issues had been neighborhood gangs and uncaring landlords.

In the midst of the planning process, CAN's president resigned, announcing that he had decided to move out of the neighborhood. Another longtime leader refused to participate, saying that he was also thinking about moving. Others, who in public at committee or commu-nity meetings would declare support for the planning process, would state privately that they had serious doubts about the value and effi-cacy of the effort.

Trying to figure out what to do, we began to turn away from thinking about grandiose strategies for coping with gentrification and toward thinking about what we could do to build CAN as an organization so that it could address gentrification. Our observations had led us to con-

clude that even if CAN had been more clear in its resolve regarding gentrification, its own internal structure was inadequate to take on the implementation of any long-term strategies we might propose.

Like many presumably viable community-based organizations, CAN was quite small, consisting of a core group of no more than 20 dedicated leaders and a dues-paying membership of no more than 200. In addition to the normal set of officers, CAN carried out its work through only two committees, steering and planning. Many CAN officers and members participated on both committees. Neither committee seemed to be capable of making decisions. Members of both committees were unsure as to what their responsibilities were and what authority they possessed. As a way of avoiding making any decisions, it was common to refer an item arising in one committee to the other committee.

CAN's leaders seemed to be uncomfortable with becoming actively involved in the process of neighborhood development. Their previous activities had been associated almost exclusively with reacting to specific neighborhood problems that had been brought to the attention of the organization by its members. For example, CAN had been successful in getting two blocks in the neighborhood rezoned from commercial designation to residential. This came about, however, only after two of CAN's leaders who lived on the blocks in question had become concerned that their homes were threatened by the possibility of commercial expansion.

The idea of developing a community plan and then becoming actively involved in helping to set the direction of neighborhood development was disturbing to many members of the organization. They expressed considerable doubt as to whether they as individuals or as a group had the intelligence, experience, and resources to carry out what were unknown tasks. They seemed to be excited and hopeful about the help they were getting from the Voorhees Center but were very reluctant to commit to implementing any of the strategies we had begun to suggest.

CAN's leaders also felt isolated from potential allies. Only the community organizer and one or two members had any links with the two aldermen who represented the community. The organizer knew about and had even been in contact with community-based development groups throughout the city and was aware of many resources that

could be used to implement any planning strategies that might be developed. However, CAN members, for the most part, were unaware of them.

## ANALYZING CAN'S WEAKNESSES

In its current state, CAN lacked the capacity to become a community development organization. Significant changes would be needed. This was the conclusion the graduate student assigned to the CAN project and I came to. We also felt that if we were to involve CAN in any plan for the neighborhood, we would have to pay as much attention to strengthening CAN as an organization as we would pay to identifying strategies for coping with gentrification. To begin thinking about how we might affect positive change in CAN, we turned to recent work (at that time) by community psychologists John Prestby and Abraham Wandersman. Their work, which fell outside of the usual frame of planning, had been recommended to me by a colleague in the psychology department at the University of Illinois at Chicago.

Prestby and Wandersman (1985) had studied 17 block club organizations in Nashville, Tennessee, to find out why some succeed but others fail. They had constructed a framework of four variables felt to be critical to an organization's success: resource acquisition, maintenance, production, and external goal attainment. We used this framework to both analyze the current structure of CAN and to devise the outline of an action plan for the organization that we hoped would help CAN become more development oriented. Our analysis of CAN's structure, using each of the four variables, is reviewed in the following paragraphs.

### Resource Acquisition

An organization needs both external and internal resources. CAN, as was previously noted, had few external links and thus had almost no external resources it could call on. Because it was a relatively small organization, its internal resources were also limited. It relied primarily on the hard work of its members and had never determined what

resources and skills it actually needed or had made any effort to obtain them.

## Organizational Maintenance

CAN was having trouble maintaining itself. The neighborhood was already changing, and CAN was having trouble retaining its limited membership and resources. Riger and Laurakas (1981) have found that resident support for community organizations depends on the maintenance of a psychological and physical attachment to a community. Neighborhood change can cause a community organization's members to feel detached from the community and to become ambivalent in their support of the organization. This, of course, weakens the organization.

Thus, when rapid changes occur, as when gentrification happens, it is difficult for an organization to maintain itself, and CAN was no exception. Some CAN leaders clearly felt estranged from their changing neighborhood and expressed it by either becoming ambivalent about continuing to support the organization and its aims or by expressing doubt as to whether the organization could actually accomplish anything in such turbulent conditions. Other longtime CAN members were being displaced as their rental units were targeted for rehabilitation and resale or their rents were raised. As they were forced out of the community, they left the organization. Remaining CAN members were also not sure whether they wanted to try to rebuild the organization by recruiting new members from residents who were moving into the community. Although they were concerned about their declining membership base of old-timers, they also worried that recruiting newcomers to the neighborhood might result in fundamental changes in the organization.

## Production

CAN had a relatively good record of being able to produce results. It had reacted to detrimental conditions in the neighborhood by fighting those things it considered to be causes of neighborhood decline, and it had its share of victories. But most often, it had sought to maintain a kind of status quo in the neighborhood and had not sought to bring about fundamental neighborhood changes.

Gentrification was putting CAN's ability to produce in jeopardy. Reaction and attempts to maintain the status quo are unlikely to be effective modes when gentrification is occurring in a neighborhood. Efforts to maintain the status quo will be overwhelmed by the market forces that are bringing about the changes. Under such conditions, a neighborhood organization needs to be more proactive if it is to produce results. Given the situation, it was becoming increasingly hard for CAN to accomplish much of anything, even in the area of internal production, such as recruiting new members and raising funds.

## Goal Attainment

CAN had neither a mission statement nor any formally articulated set of goals. As noted, its previous efforts had mostly been ad hoc actions taken under a rather vague implicit goal of neighborhood improvement. Neighborhood improvement no longer seemed to be an appropriate goal because the neighborhood was in fact "improving" but in a way that members felt was undesirable. But there was little consensus within the organization as to what its goals should be. There was, however, a consensus that unless the members could agree to coalesce around a shared vision for the neighborhood, it would be difficult, perhaps impossible, for CAN to become an active participant in helping to shape the future of the community.

# A PLAN FOR THE COMMUNITY
# AND THE ORGANIZATION

We used the four Prestby and Wandersman (1985) variables to develop a set of strategies intended to build organizational capacity and to move CAN toward a development role. Our ideas were presented on a Saturday morning in early March 1986 during a workshop with members of both of CAN's standing committees. After we explained each of the four variables and how we felt they related to CAN, we worked with the members to identify a set of realistic objectives specifically related to each variable in the framework. When the day was finished, we had created a short-term plan for the organization. It contained the following objectives.

## Resource Acquisition

1. Recruit and train at least 12 new members to take on leadership positions
2. Appoint specific individuals to act as liaisons to other organizations active in the neighborhood
3. Create a committee to consider whether CAN should drop its current affiliation with the Lakeview Citizens Council and affiliate instead with the Lincoln Park Conservation Association, the umbrella organization covering the rest of the Lincoln Park community area
4. Seek greater support from institutions both inside and outside the community

## Maintenance

1. Reaffirm the organization's "open membership and equal representation policy"
2. Restructure the organization to clarify roles and responsibilities for the purpose of increasing responsiveness to changing neighborhood conditions
3. Redesign the organization's brochure and develop a regular schedule for its newsletter
4. Raise funds to hire a new staff person and purchase a computer

## Production

1. Double the membership of the organization
2. Initiate monthly issue forums to take the "pulse" of the community
3. Develop policies that reflect the organization's position on neighborhood development
4. Initiate monitoring of housing and industrial changes in the neighborhood
5. Create a housing resource center to assist exiting homeowners, apartment owners, and renters
6. Create an "image enhancement" program

## Goal Attainment

1. Accomplish each of the objectives in the first three categories

We left the meeting with a good feeling. CAN now had a plan and strategies that, if successfully carried out, would give it a voice in determining the future direction of the neighborhood.

CAN immediately added the plan to a larger proposal it was submitting to the city of Chicago to obtain community development block grant funds for doing community planning. In a few weeks, we learned that the proposal had been successful and that CAN would receive a $17,000 block grant allocation. It appeared as though CAN had a real chance to move from being a small reactive neighborhood association to becoming a proactive community development organization.

## CAN CHANGES WITH THE NEIGHBORHOOD AND THE PROJECT ENDS

Although CAN was presumably beginning to implement the strategies for building membership, we turned our attention to doing a background study for the creation of the housing resource center, the fifth objective under the goal of production in the plan. This objective, we believed, would be the key to CAN's becoming more actively involved in the changes occurring in the neighborhood. We thought of the "housing center" as sort of a transitional step between the organization as it now existed and a full-fledged community development corporation (CDC). We envisioned the possible activities of the center to be things such as managing a mini-loan pool for home improvement projects, providing referrals for larger loans through existing city programs and private lenders, and providing assistance to renters seeking housing.

Almost immediately we began to encounter problems. I received a phone call from CAN's community organizer informing me that he was leaving the organization to enroll in a graduate urban planning program in another state. Although I was happy for him, I was also concerned because he had been our primary link to CAN and was our strongest supporter. At his suggestion, I contacted the director of Christopher House, who would be responsible for replacing the organizer because the contribution of Christopher House to CAN was the underwriting of the organizer's salary.

The director was new to the West DePaul community, having been at Christopher House just about a month. He said that Christopher House was in the process of rethinking its priorities in the neighborhood and suggested I attend a meeting with him and a few other people he was calling together to do some organizational planning.

I went to the meeting hoping to get a commitment from Christopher House to replace the community organizer. The meeting, however, concentrated primarily on how Christopher House could better meet the needs of the new families that were moving into the neighborhood. He wanted to know whether the Voorhees Center could provide planning resources to assist him in establishing new programs. Rather than discussing problems associated with encroaching gentrification, most of the meeting was spent discussing topics such as day care and after-school enrichment programs for the new families moving into the neighborhood. This seemed to be an odd discussion to be having at a settlement house, which I had always thought attempted to serve the poor and disadvantaged. At the end of the meeting, the director said he was unsure as to whether to continue support for CAN and that he thought it unlikely that Christopher House would hire another organizer.

My graduate assistant was also having trouble making contact with CAN's planning committee. Lacking the organizer, she was attempting to work through one of CAN's leaders who had previously been supportive of our work and with whom the student felt comfortable. Missed and unreturned telephone calls became commonplace. When the student finally did reach the leader, she was told that the committee was not ready to meet with us and she would get back to us later with a meeting time and place.

Finally in mid-summer, we insisted on meeting with the planning group. The student had done considerable work and had prepared a portfolio of information regarding the structure of a housing resource center and felt she could go no further without consulting with the planning committee. In addition, both she and I were feeling uneasy about the situation and were concerned that the project was stalled.

Arriving at Christopher House and entering the room set aside for the meeting, we sensed our efforts were in trouble. We recognized only about half of the 15 or so people in attendance. The rest were complete strangers to us, and we later found out that they had joined CAN since we had last met with the organization. The most outspoken committee member, an unfamiliar face, introduced herself as a new real estate agent in the community.

The meeting began, and it soon became clear that the focus of CAN had shifted. Much of the discussion was about the positive "improvements" that were occurring in the neighborhood and how CAN could support the "improvements" and assist in attracting more newcomers to the neighborhood. Even those committee members we knew and who had voiced support for developing strategies to limit the negative impacts of gentrification seemed to have shifted their positions and had become supporters of gentrification.

We were finally given an opportunity to make our presentation. The student laid out our idea of assisting homeowners and renters to remain in the neighborhood through the creation of a housing resource center, outlined what we thought the center should look like, and then asked the group for their comments and questions. The basic response was to question why anything was needed to help the existing owners and renters. Shouldn't we be thinking of ways to help people buy into the neighborhood? Existing owners don't need help, and wouldn't the neighborhood be better off without so many renters, particularly low-income renters? Anyway, as rising prices and rents were sure to drive out low-income renters, why should efforts be made to assist them because they will soon be gone anyway? Needless to say, these were not the kind of comments and questions we had expected.

The real estate agent, turning to the group, asked why CAN had hired the Voorhees Center in the first place, as she could not see how what we were doing was of any interest to the organization. No one, not even our presumed friends, came to our rescue. After a short discussion about the purposes of having our assistance, a vote was taken to dismiss us. It passed with an overwhelming majority. We excused ourselves, left the meeting, and no longer had either a client or a project.

Today, West DePaul is a fully gentrified community. New town houses have been built on vacant lots, and nearly every existing house has been remodeled. Some of the old factory buildings, which had provided jobs for the neighborhood residents, have been torn down and replaced by town house developments, and others have been converted into expensive loft condominiums. The rate of change was rapid and complete.[6] It took only 2 to 3 years for the neighborhood to completely change its character.

Returning to Table 5.1, we can see how the community changed demographically between 1980 and 1990. All four of the census tracts that make up the neighborhood show major shifts resulting from gentrification. Even in the western-most census tract, where the impacts have been the least, home prices rose nearly $50,000 above the citywide average, even though rents remained low.[7] The median house price in Tract 705, where some gentrification was no doubt occurring in 1980, rose from $53,900, which was just above the citywide median, to $247,200, an increase of approximately 4.50 times. This compares with an increase in the citywide median of only 1.6 times.

## THE FAILURE OF CAN TO CONFRONT GENTRIFICATION

David Knoke and James Woods (1981) assert that voluntary community organizations are most effective when the environment in which they operate is stable and relatively simple, when opportunities to mobilize their membership are frequent, and when they are well integrated into the community. The onset of gentrification destabilizes the environment and makes it more complex. Yet it also provides an opportunity for an organization to mobilize its membership. Whether it can successfully mobilize its members is probably the test of how well it really represents or is integrated into the community.

A community organization may not be sure how it should react to gentrification. Trying to stop it completely may prove futile, and over time the organization's influence is likely to wane. But being too accepting of "newcomers" may also weaken an organization if old members are offended and leave. Not only can the organization lose members, but it also may ultimately lose its longtime leaders and its history as an organization. Walking the line between being totally opposed and totally accepting of gentrification is no doubt a difficult task but one that CAN initially seemed prepared to attempt.

Henig's (1981) study of community organizations in the gentrifying Adams Morgan neighborhood in Washington, D.C., suggests that successfully navigating a middle position is difficult and possibly impossible. Caught up in the rapid changes that gentrification brings, community groups are more likely to be followers reacting to change rather than leaders attempting to direct it. Of three organizations existing in Adams Morgan at the onset of gentrification, one fought it, and the other two gave unqualified support to it. Gentrification eventually changed the neighborhood completely, but Henig argues that none of the three organizations benefited from either their stance or their actions.

Based on what had seemed to be the desire of CAN's leaders to attempt to have some control over the gentrification of the neighborhood, we had proposed a strategy recognizing that gentrification was occurring and that big neighborhood changes were sure to happen, but it was hoped that they would allow for some maintenance of the existing neighborhood and its residents. But we never got a chance to put the strategy into action. CAN was weakened initially as several members and one or two key leaders moved from the neighborhood. The untimely death of yet another leader further weakened the organization. The remaining leaders sought out new members, hoping to keep the organization viable. But the new members brought with them a different set of goals and objectives and essentially changed the organization. When we met for the final time with CAN, it no longer wished to navigate toward some middle ground. It now embraced gentrification. Whether CAN could have done anything to have stopped or even slowed gentrification is questionable, but the answer will never be known.

*     *     *

Although we were initially disappointed with our failure to assist CAN in implementing strategies that would at least ameliorate the impacts of gentrification, in retrospect there had never been much of a prospect for success. CAN had certainly waited too long to act, and by the time we were asked to assist them, it was no doubt much too late for any strategy to be effective.

But it was not simply timing that caused the planning effort to be of little use. CAN organizationally was unable to respond at the level that would have been necessary if it were to have been effective. It was too small, too poorly organized, and acting almost alone. It was unable to internally focus on the issue of gentrification and to externally obtain resources that it would need to develop the programs we would be suggesting. Even as we developed our plan, we knew that moving CAN from a "neighborhood association" to a more community development mode would be difficult.

Our efforts, although well meaning, could have succeeded only if CAN itself had moved toward greater internal organization. Instead, it first collapsed internally, then reconstituted itself as an almost new organization, one that favored gentrification. As a result, we were left without a client, and our plan was never really considered.

## NOTES

1. Hirsch (1983), for example, describes urban renewal efforts on Chicago's south side from their beginnings in the 1940s through the development of high-rise public housing in the early 1960s. It follows from Hirsch's work that the current gentrification that is occurring in the Kenwood community and further north in the Grand Boulevard community, which is increasingly being referred to as "Bronzeville," is a continuation of the goals and objectives of civic leaders that were well articulated by the 1940s.

2. Chicago is officially divided into 77 community areas with fixed boundaries. These areas were first identified by pioneering urban sociologists at the University of Chicago in the 1920s. Since 1930, census data have been published by community area in what is known as the *Local Community Fact Book* (Chicago Fact Book Consortium, 1995). The original number of community areas was 75, but O'Hare Airport was later added as community area 76, and in the early 1980s the Uptown area was split in two, creating a 77th area known as Edgewater.

3. The Old Town area became known for its bars and nightlife. The nationally known comedy troop, Second City, had and continues to have its home in the Old Town area.

4. As the 1980s continued, gentrification moved northward from Lincoln Park into Lakeview. Today Lakeview, like its neighbor to the south, consists of expensive apartments, condominiums and town houses, and trendy stores and restaurants.

5. The organization, however, was not fully representative of the neighborhood. Homeowners were overrepresented in leadership positions, and blacks and Latinos were underrepresented.

6. About a year after our project ended, I was accompanying a group of students from Albion College who were in Chicago on a field trip as part of their urban politics class. We visited West DePaul, and what we noticed most were the sounds of hammers and saws. It appeared to us that nearly every house in the neighborhood was undergoing some kind of renovation.

7. In the westernmost census tract between 1980 and 1990, there was a doubling of the percentage of the population with some college education, although median family income was still quite low.

# 6

# THE SPORTS STADIUM VERSUS
# THE NEIGHBORHOOD

The story of the second case study begins more than 50 years ago when the U.S. Public Housing Authority built Wentworth Gardens, 420 units of public housing, at the southern end of Chicago's Armour Square community area between 37th and 39th Streets (see Figure 6.1). Construction was started in 1945, and the housing was intended for black war workers, but World War II ended before the development was ready for occupancy.[1] So it was decided that the units would be used as low-rent housing, and the new development—37 buildings consisting of two-story row houses with a cluster of three-story apartment buildings in the center—became part of the expanding inventory of the then new Chicago Housing Authority (CHA)[2] (Bowley, 1978). From the very beginning, Wentworth Gardens housed black families.

With the building of Wentworth Gardens, 35th Street came to be a dividing line in Armour Square. As blacks moved into the public housing, whites, who lived in the homes between the development and 35th

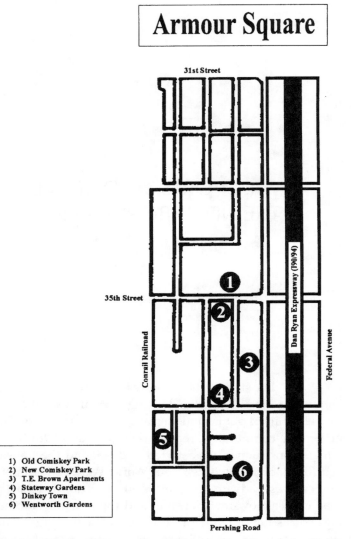

**Figure 6.1.** Map of the South Armour Square Neighborhood Boundaries: 31st Street, Federal, Pershing (39th Street), and Conrail (RR)

Street, moved out and were replaced by black homeowners and renters. North of 35th Street, the neighborhood remained white, and blacks did not dare cross 35th for fear of being attacked and beaten. One longtime resident who owned a home south of 35th Street told me that shortly after she moved into the neighborhood, she had naively taken her small child in a baby buggy for a walk in Armour Square Park, located at about 34th Street. But her walk was short-lived when she was confronted by a group of men who chased her and her baby back across 35th Street.[3]

From the beginning, Wentworth Gardens has been one of CHA's more desirable developments. Architecturally, the units are somewhat nondescript, but the development has a relatively low density, and there is considerable open space between buildings. The row houses and three-story buildings stand in stark contrast to the crime and violence that plagued the Robert Taylor and Stateway Garden high-rises located only a few hundred yards to the east. Even though Wentworth is close to the infamous high rises, its residents have virtually no contact with their CHA neighbors. The developments are physically separated by the multilane Dan Ryan Expressway, which cuts a wide swath through the south side of Chicago.

Over time, the area south of 35th became known as South Armour Square, and it became the kind of a place that sociologists refer to as a defended neighborhood (Suttles, 1972). It was a small community of African Americans, more or less cut off from the rest of the city. There was a sharp neighborhood boundary at 35th Street, separating South Armour Square from the white neighborhood to its north and another sharp boundary on the neighborhood's west side along an elevated railroad embankment, separating it from the all-white community area of Bridgeport (see Figure 6.1). The construction of the Dan Ryan Expressway in the 1960s created another sharp boundary to the east and resulted in a further isolation of the South Armour Square neighborhood.

Since 1912, long before Wentworth Gardens was built and African Americans had moved into South Armour Square, Comiskey Park, the home of the Chicago White Sox American League baseball team, had been a community fixture. Comiskey Park sat just north of 35th Street at Shields, only 2 blocks from the public housing and just across the street from the homes of South Armour Square. Hardly a "gem" of a stadium because it had been built at a modest cost and reflected the Sox's original owner, Charles Comiskey, and his justified reputation as a "penny pincher," Comiskey Park nonetheless had a colorful history.

Over the years, Comiskey Park had seen good times and bad. Occasionally, the team would win a league pennant, the last one being in 1959. Perhaps, though, the franchise was best known for its infamous 1919 squad, known as the "Black Sox," because several players had conspired with gamblers to purposely lose the World Series. But besides the Sox, many famous American League stars had played at Comiskey, including Babe Ruth, who as legend has it, would dash across 35th Street between innings for a beer and a hot dog at McCuddies, a bar that still sat across from the park in the mid-1980s.

## THE NEIGHBORHOOD OF
## SOUTH ARMOUR SQUARE

By 1980, South Armour Square was home to nearly 2,300 individuals, all but 7 of whom were black (see Figure 6.2 and Table 6.1). Wentworth Gardens accounted for just about two thirds of the housing units, the rest being divided between a mid-rise federally subsidized senior and handicapped building, T. E. Brown Apartments,[4] which contained 116 units, and approximately 178 single-family homes or two flats. Approximately 70 of the single-family and two-flat units were owner occupied. There had been some loss of housing between 1970 and 1980, primarily in the area known to South Armour residents as "dinkey town," which was west of Wentworth Gardens and south of 37th Street. North of 37th there were scattered vacant lots where housing had once stood, but most of the lots still contained homes that, although old, were well maintained.

South Armour Square had the reputation of being a close-knit community, although in reality it consisted of several small "street" neighborhood areas, and there was very little interaction between people from different parts of the community.[5] This was particularly true with respect to contact between the people who lived in the area between 35th and 37th Streets and the residents of Wentworth Gardens. There were conditions, however, under which the area functioned as a community. Children from all of South Armour Square attended the Robert S. Abbott elementary school at the corner of Wells and 37th, and many residents from the various parts of the community attended the Progressive Baptist Church at the corner of Wentworth and 37th. There

**TABLE 6.1** Demographics of the South Armour Square Neighborhood

|  | Census Tract | |
|---|---|---|
|  | *3405* | *3406* |
| **1980** | | |
| Total population | 1,702 | 2,233 |
| % black | 25 | 100 |
| % of Spanish origin | 2 | 0 |
| % 13 years old and younger | 12 | 32 |
| % 65 years old and older | 34 | 5 |
| Median family income, 1979 | $16,957 | $9,052 |
| % income below poverty level | 20 | 45 |
| % white-collar workers | 54 | 38 |
| Population per household | 1.8 | 3.9 |
| | | |
| Total housing units | 948 | 634 |
| % condominiums | 0 | 0 |
| % built 1970 or later | 0.15 | 1 |
| % owner occupied | 0.18 | 11 |
| Median value: Owner units | $30,000 | $16,300 |
| Median rent: Rental units | $77 | $105 |
| **1990** | | |
| Total population | 1,477 | 1,479 |
| % black | 30 | 100 |
| % of Spanish origin | 4 | 0 |
| % 13 years old and younger | 8 | 32 |
| % 65 years old and older | 38 | 8 |
| Median family income, 1989 | $32,366 | $5,652 |
| % income below poverty level | 7 | 88 |
| % white-collar workers | 55 | 45 |
| Population per household | 1.7 | 2.8 |
| | | |
| Total housing units | 914 | 536 |
| % condominiums | 0 | 0 |
| % built 1980 or later | 2 | 26 |
| % owner occupied | 19 | 0 |
| Median value: Owner units | $70,500 | 0.00 |
| Median rent: Rental units | $237 | $135 |

NOTE: Data in this table are from the U.S. census, as organized and reported in the *Local Community Fact Book, Chicago Metropolitan Area 1990* (Chicago Fact Book Consortium, 1995).

were essentially no other places where people could interact other than at a few small variety stores that existed primarily for the selling of liquor.

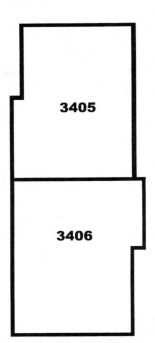

**Figure 6.1.** Map of the South Armour Square Neighborhood Boundaries: 31st Street, Federal, Pershing (39th Street), and Conrail (RR)

Wentworth Gardens, however, within its own boundaries, was a very active community. As early as 1968, the then newly formed Wentworth Gardens Local Advisory Council[6] (LAC) had sought and received permission from the CHA to open a coin laundry in the basement of one of the three-story apartment buildings. The LAC recruited residents from the development to staff the laundry and used the proceeds for community programs, youth scholarships, and Christmas gift certificates. The coin laundry became a fixture of the development and has been in continuous operation now for more than 30 years.

In 1973, the LAC added another economic development and community service activity, a small grocery store, intended to offset the lack of shopping opportunities in the nearby community. The proceeds from the grocery store also have been used to underwrite community programs. These programs, especially a summer youth program, are planned and run by community residents with the help of area social service agencies and have been quite successful. The key to their success has been the existence of a network of interested residents and a strong sense of community.

With the passage of time, South Armour Square grew old. Wentworth Gardens suffered from the same neglect that afflicted all CHA developments, and the modest houses north of 37th Street just simply aged. A few fell into total disrepair and were demolished, but most of the houses, although weather-beaten, remained well tended (Joravsky, 1988). According to attorney Jim Chapman, nobody would mistake South Armour for Chicago's Gold Coast, but "it was a good little neighborhood" (Wiltz, 1990).

## PROPOSALS FOR A NEW COMISKEY PARK
## THREATEN SOUTH ARMOUR SQUARE

By 1980, the stage was being set for massive change in South Armour Square. After the 1980 baseball season, the owner of the White Sox, the legendary Bill Veeck,[7] sold the team to a consortium of investors headed by two businessmen, Eddie Einhorn and Jerry Reinsdorf. The franchise the two inherited was nearly bankrupt (Joravsky, 1988).

As part of their plan to rebuild the White Sox, Einhorn and Reinsdorf announced that they needed a new stadium, and they began looking for a governmental entity that would build one for them. To justify a new stadium, they claimed that the existing Comiskey Park was structurally unsafe and would be too costly to repair.

They also began suggesting that unless they got a new stadium in the Chicago area, they might be forced to move the team to another city. Several Florida cities expressed interest, and St. Petersburg actually began construction of a new facility, partially at least with hopes that they might attract the Sox. There was also considerable talk that the team might move to Chicago's suburbs, but this possibility ended when in November 1986, voters of Addison, the site that appeared to be the choice of the Sox's owners, rejected by referendum the Sox's overtures.[8]

Harold Washington, the first black mayor to be elected in Chicago, did not want to lose the White Sox. Following his election in 1983, Chicago's business community had expressed considerable doubt whether the mayor would be interested in their concerns, and they warned anyone who would listen that if the mayor ignored them, the city was sure to suffer. Not wanting to be the mayor who presided over the disinvestment of Chicago and fully aware of what had happened in other midwestern cities, such as Detroit and Gary, after black mayors had been elected, Washington was determined to keep the city's business climate strong.

Although the loss of the Sox to the suburbs or to Florida would not have had that great of an impact on Chicago's economy, it would have been seen as a sign of Washington's weakness with respect to the business community. Thus, Washington placed a high priority on retaining the White Sox, and he appointed his commissioner of economic development, Rob Mier, to head up negotiations between the city, the state, and the White Sox.

The city moved quickly in response to the White Sox's threats, and shortly after the voters in Addison rejected the Sox's move to the suburbs, it was announced that the city, state, and the White Sox had formed an "alliance" to build a new stadium. The alliance was cemented in late 1986, when the state's General Assembly passed enabling legislation permitting the creation of an Illinois Sports Facilities Authority, a quasi-governmental agency that would oversee the construction of a new ballpark for the White Sox. Among the powers the new authority was to have would be the power of "quick take," meaning it could condemn property, take it over, and later settle on a payment in court.

Residents of South Armour Square soon discovered that their neighborhood was the target for the new stadium. In January 1987, Mayor Washington accompanied Governor James Thompson to Comiskey Park for the signing of the Sports Facilities Authority legislation, and residents read in the newspapers that the new park was going to be built south of 35th Street instead of on or near the existing site north of 35th. The threat to the neighborhood became even clearer in March when the *Chicago Sun-Times* published a map showing the location of the new stadium. According to the map, the stadium, as well as the parking lot surrounding it, was to cover all of the area between 35th and 37th Streets and about half of Wentworth Gardens south of 37th. Although the neighborhood, most of it at least, was to be destroyed, as of yet not one single official of the city, state, or the White Sox had made any effort to talk with the residents of South Armour Square.

## THE FIGHT TO SAVE SOUTH ARMOUR SQUARE

It was about this time, in the winter of 1987, that a few residents of Wentworth Gardens approached and sought assistance from Sheila Radford-Hill, a community organizer with the Chicago Area Project.[9] Realizing that there were no formal community organizations in South Armour Square and thus no organized group ready to fight for the survival of the community, Radford-Hill immediately set out to create a structure that would bring together all of the community, including the homeowners, renters, residents of T. E. Brown Apartments, and residents of Wentworth Gardens. The organization that emerged became known as the South Armour Square Neighborhood Coalition (SASNC).

At the same time, Radford-Hill also recognized that there was no organization in Wentworth Gardens that could become part of the newly formed coalition. Although there was an active LAC, it could not join the coalition out of fear that by taking an activist stand against the stadium, it might jeopardize its relationship with the Chicago Housing Authority. So a second organization, Wentworth Residents United for Survival (WRUS), was formed to create a focal point around which the public housing residents could organize.

My first contact with the SASNC and WRUS occurred in April 1987 at a conference at the University of Illinois at Chicago. Both Sheila Radford-Hill and I had been asked to speak on a panel about public housing. Radford-Hill had brought two South Armour residents with her who also spoke. The four of us chatted briefly at the end of the panel, and I offered my assistance and that of the Voorhees Center to helping the community in its fight to save the neighborhood. About a week later, Radford-Hill called to accept my offer.

After a meeting with Radford-Hill and a few resident leaders from the coalition, it was agreed that the Voorhees Center would begin working with the coalition to develop a neighborhood plan, which would serve as an alternative to the city's plan to destroy the neighborhood by building the stadium. Radford-Hill felt that creating a vision for South Armour's future would complement her organizing work, which was focusing on stopping the construction of the stadium. A neighborhood plan, she believed, would give residents hope that the neighborhood could not only survive but also be improved in the long run.

The planning process started a few weeks later on a Saturday afternoon in the community room of the T. E. Brown Apartments. Approximately 40 residents, representing all areas of South Armour Square, attended and participated. George Marshall, a homeowner who had been elected president of the coalition, was there as were Mrs. Dorothy Driver and Mrs. Hallie Amey, both longtime resident leaders at Wentworth Gardens. After I presented the idea of doing a plan and outlining how a community-based plan could be created, we started by making lists of things residents would like to see happening in South Armour Square. Along with a general cleanup and improvement of the streets and vacant lots, the residents identified needs for a basic retail complex that would minimally house a grocery store and a drug store; some kind of a medical facility, probably a satellite outpatient site of a local hospital; and a small neighborhood park. None of the residents present at the meeting had ever been seriously asked about what the neighborhood's needs were, and as the meeting progressed, they became excited about the possibility of developing their own neighborhood vision.

We met twice more at the T. E. Brown community room. At the second meeting, we used wall-sized maps of South Armour Square to sketch out a land use plan that included the elements that had been identified at the previous meeting. There was lively discussion over many issues, such as just where to put the park so it would serve both the public housing residents and the homeowners and renters. People from north and south of 37th who had not known each other just a month or two ago became engaged in the process of developing a plan that would benefit the whole community and were working together as though they had been friends all their lives. In the end, there was general agreement that most of the needed facilities should be located in the much-deteriorated "dinkey town" area of the community because this would achieve dual goals of cleaning up a blighted section of the neighborhood and creating needed facilities that would be accessible to all South Armour Square residents.

At the third meeting, my students and I returned with completed land use maps showing what had been accomplished during the previous two meetings. After Sheila Radford-Hill spoke to the group about the need to carry the plan forward and to use it as a tool in the struggle to preserve the neighborhood, the plan was ratified by voice vote. This was followed by a "salad lunch" potluck that served as a celebration of what had been accomplished and an opportunity to further cement relationships across internal neighborhood boundaries.

Meanwhile, very little official progress was being made on the stadium proposal, and this resulted in giving hope to the residents that they might actually be able to stop or at least significantly alter the stadium plans. Feuding between the mayor and the governor over the structure and makeup of the Sports Facilities Authority lasted well into the summer of 1987, and it was fall before a chair and members of the authority's board of directors were appointed and the authority actually began to function. During this period, the White Sox proceeded to negotiate an agreement with the city of St. Petersburg, and for a while, it looked as though they would actually be moving to that Florida city. Lack of progress toward building the stadium made it appear that the neighborhood plan might not be an exercise in futility and that maybe it might someday actually be implemented.

On the other hand, progress was being made with the community organizing effort. The South Armour Square coalition was meeting regularly and holding frequent rallies and demonstrations. Despite early assurances from Mayor Washington that residents interpreted as a promise not to demolish their homes[10] (Joravsky, 1987), news reports continued to state that homes would be demolished to make way for the stadium. So residents continued to put pressure on the Washington administration by holding rallies outside of the existing Comiskey Park and at City Hall. They would later hold demonstrations that disrupted Sports Facilities Authority meetings after it was formed and had begun to meet. The pressure was apparently making the city administration uncomfortable. Rob Mier, the mayor's point man on the stadium issue, was quoted as saying that the residents were being exploited by fear-mongering activists and troublemakers (Joravsky,

1987). To this assertion, Radford-Hill responded, "That's par for the course. That's the first thing people in power say about organizers" (Joravsky, 1987, p. 17).

The strategy adopted by the coalition was not to try and keep the stadium from being built but rather to have it built somewhere else, preferably north of 35th on the present Comiskey Park site or next to it. This tactic garnered them the support of a small but influential group of 40 or so White Sox fans and historic preservationists who had organized into a group known as Save Our Sox (SOX). The SOX group actually wanted to see the existing park renovated but was willing to support the efforts of the South Armour Square coalition. A Save Our Sox member, Mary O'Connell, was the editor of *The Neighborhood Works*, a respected magazine published by Chicago's Center for Neighborhood Technology, and she was particularly helpful in getting both the SOX's and South Armour Square coalition's message to the general public.

There was also additional support. A local architect, Philip Bass, had developed his own plan for a more neighborhood-friendly stadium to be located north of 35th, and his proposal was publicized in the local press.[11] Eventually, the *Chicago Sun-Times* ("Save the Sox," 1988) threw some support in favor of preserving South Armour Square by suggesting that the South Armour residents should be left alone and that the stadium should be built instead on Chicago's near west side.

The coalition decided to move ahead with the community plan and to hold a conference in the fall that would act as both a kickoff for the plan's implementation and put the city on notice that it was serious about not only preserving the neighborhood but also improving it. In preparing for the conference, I had a student undertake a thorough external survey of all residential structures in South Armour Square, excluding the buildings at Wentworth Gardens. She reported that of the existing 81 structures, only 4 were vacant, and all but 25 were in sound conditions. Twenty structures, however, were found to be in need of significant structural repairs. Our assessment of the current conditions suggested to us that renovation would be possible.

The conference, billed as a "neighborhood improvement conference," was held November 12 at the Center for Inner-City Studies of Northeastern Illinois University, a site about a mile from South Armour Square. More than 100 residents were taken to the center by bus to attend the daylong conference. Speakers at a morning plenary session, during which the neighborhood plan was introduced, included Maria Choca from the city's Department of Planning[12] as well as representatives from several Chicago-area community development corporations and community organizations. Specific "how-to" sessions followed to instruct community residents in skills relating to the following:

1. researching and analyzing land use patterns in communities,
2. marketing a community,
3. empowering residents of public housing, and
4. developing leadership in organizations.

The last two sessions were targeted at specific groups of individuals in attendance. Residents from Wentworth Gardens were encouraged to attend the empowerment session, which was presented by Stanley Horn, who had been heavily involved in the recent conversion of the CHA's Leclaire Courts development to resident management (which is the subject of the case study in the next chapter). Board members of the South Armour Square coalition attended the session on leadership development where they listened to several speakers.

As 1987 came to a close, it appeared that the South Armour Square coalition had made real progress. There was reason to believe that the coalition might prevail and the neighborhood might be saved.

## THE BATTLE IS LOST

As 1988 unfolded, the struggle continued. The Sports Facilities Authority began its negotiations with the White Sox over a stadium lease, and the South Armour Square coalition continued its protests. The community's hopes were raised in April, when a report, based on a study by the engineering firm George A. Kennedy and Associates, was released. It contained the finding that the existing Comiskey Park was in "relatively good shape . . . and could be renovated for perhaps less money than it would take to build a new stadium" (Spielman, 1988, p. 8). This was particularly good news to the Save Our Sox group, but it also gave new ammunition to the coalition.

But unknown to coalition members, their organizer (Sheila Radford-Hill), and the rest of us who had been providing assistance to the coalition, the city and the Sports Facilities Authority had begun secret negotiations with George Marshall, the coalition president, and with Mary Milano, an attorney who had been assisting the coalition. The purpose of these negotiations was to reach an agreement between the homeowners of South Armour and the Sports Facilities Authority regarding relocation. The authority was interested in dealing only with the homeowners. It was a classic case of an attempt to split the coalition by playing to the interests of one segment. Unfortunately, the divide-and-conquer strategy succeeded.

On August 3, Marshall appeared at a board meeting of the coalition with a 250-page agreement between the coalition and the Sports Facilities Authority and told the members present that they should sign it at once, even before reading it. Marcella Carter, the secretary of the coalition and a resident of Wentworth Gardens, immediately moved to adjourn the meeting and to reconvene in 30 days, at which time a vote would be taken. The motion passed by a vote of 10 to 3.

But the Sports Facilities Authority board ignored the vote of the coalition board and immediately began negotiating with homeowners on a one-to-one basis. The coalition was irreparably split and eventually reconstituted itself with only residents of the T. E. Brown Apartments and Wentworth Gardens as members. South Armour Square had lost the battle, and the White Sox would get their new stadium where they wanted it—on land south of 35th Street.

It would take many more months before any of the South Armour homeowners would see the land deal finalized. A final agreement was not reached until November 1989, and then it involved only 18 families and promised the construction of 14 two-family and 4 single-family replacement homes. These homes were not built until the summer of 1991.

Because the final design of the park did not involve taking any of Wentworth Gardens or the T. E. Brown property, there were no arrangements made with either group. This particularly angered residents at T. E. Brown because the park's center field exploding scoreboard ended up being only a few hundred feet from the apartment building, a continuing nuisance to the elderly residents of the development.

Even though T. E. Brown and Wentworth were spared, the reconstituted coalition filed a lawsuit in February 1989 against the city, the

Sports Facilities Authority, and the White Sox, alleging discrimination[13] in the placement of the stadium and seeking damages for the destruction of the community. It was the hope of the remaining coalition members that compensation resulting from a positive verdict could be used to implement the community revitalization plan. The case slowly proceeded through the legal system, but it was finally dismissed in the late 1990s, ending the hopes and dreams of the South Armour Square Neighborhood Coalition.

## BUT COMMUNITY DEVELOPMENT CONTINUES

The story of South Armour Square does not completely end with the building of the new Comiskey Park, although it does shift exclusively to Wentworth Gardens. At the same time the South Armour Square coalition had been formed, Wentworth Residents United for Survival (WRUS) also had formed. This organization would carry on the effort to improve the neighborhood, albeit only the public housing neighborhood.

Following the November 1987 neighborhood improvement conference and after being inspired by Stanley Horn's story of how resident management had empowered residents at Leclaire Courts, the board of WRUS decided to explore the possibility of initiating resident management at Wentworth. I agreed to provide a 10-week empowerment training program for WRUS members. To do this, I put together a team consisting of two students from the Voorhees Center, as well as Stanley Horn and Irene Johnson from Leclaire Courts. Meeting once a week between February and April in a basement room at Wentworth, we presented residents with the details and requirements of mounting a resident management effort. Our goal was to give the residents the information they would need to decide whether they, as a group, wanted to work toward resident management.

The consternation resulting from the split in the South Armour Square coalition occupied much of the energy available during the summer and fall of 1988, so it took until February 1989 for the members of WRUS and the Wentworth Gardens LAC to formally vote to create a resident management corporation and to seek CHA and Department of Housing and Urban Development (HUD) assistance in formal resident management training. By this time, two other organizations had been recruited to assist the residents. Besides the Voorhees Center, there was the Lindemann Center from Northern Illinois University, which would assist the residents in negotiating the maze of HUD and CHA regulations pertaining to resident management, and the Center for Urban Economic Development (UICUED) at the University of Illinois at Chicago, which would assist in economic development planning.

Pat Wright of UICUED would become a particularly important adviser to the residents. Wright had first become involved with Wentworth residents at the November 1987 planning conference, and she would continue working with them for many years. When I left the directorship of the Voorhees Center in 1990, she became its new director. Since then, Wright, along with Roberta Feldman, an architect at the School of Architecture and Community Design Center at the University of Illinois at Chicago, and Susan Stahl, a sociologist at Northeastern Illinois University, have provided the bulk of the assistance to Wentworth residents.

In December 1989, WRUS formally changed its name to the Wentworth Gardens Resident Management Corporation, and a board of directors was officially installed a month later in January 1990. At the request of the new board, I worked with Sheila Radford-Hill[14] to deliver a second round of resident empowerment training between April and October 1990. In June, a development-wide referendum was held on whether Wentworth Gardens should be resident managed. It passed by a vote of 107 to 4.

Since then, many good things have happened at Wentworth Gardens. Most of the buildings have received new windows and doors, and leaking roofs have been repaired. The old asphalt "park" in the center of the development was torn up and replaced with grass and modern playground equipment. New community gardens have been started, and spirits have been raised. But the new economic development has not happened due to a lack of financial resources. And full resident management of Wentworth Gardens, after what seemed to be years during which the CHA appeared to be disinterested, was finally accomplished during the summer of 1998.[15]

*   *   *

On January 17, 1989, a memorial service was held at the T. E. Brown Apartments for the South Armour Square neighborhood. Wentworth Gardens and T. E. Brown residents came together to officially recognize that the fight to save the neighborhood had been lost. But a memorial service is more than a time to mourn. It is also a time to look forward, and those in attendance used this event as an opportunity to reaffirm their commitment to neighborhood development.

It was probably unrealistic to have believed that the residents of South Armour Square could have ever won out over the powerful forces of the city, the state, and the Chicago White Sox. From the beginning, none of these three entities showed any interest in compromising or accommodating any of the neighborhood's wishes. The White Sox wanted a new baseball park, and the city and state were determined to see that they got what they wanted.[16]

But the effort to build a coalition of homeowners, senior citizens, and public housing residents that would fight the stadium was also flawed. Perhaps there had been too little time to bring together people who, before the threat of the stadium, had not known each other. The homeowners of South Armour Square had never viewed the Wentworth residents as their neighbors, and it appears that they were only willing to join with them to advance their personal agendas and not to save the community. When the opportunity arose to arrange a deal that seemed beneficial, the homeowners did not hesitate to abandon their newly formed friendships. As has been pointed out by Rubin and Rubin (1992), neighborhood allegiances are never very strong to begin with, and the homeowners demonstrated this by choosing personal gain over neighborhood solidarity.

But some good did come out of the loss of the stadium struggle. Community involvement at Wentworth Gardens was reinvigorated, and residents began to seek out resident management as a way of gaining more control over their public housing community. Although it would take years before they would achieve this new goal, resident efforts would bring benefits to the community. In a sense, the Wentworth residents lost the war but lived to fight again.

## NOTES

1. By federal ruling, public housing could not be built during the war unless it was meant for war-industry workers. During the war, there was a significant migration of blacks from the south seeking jobs in industries of the city. Within a short time, the city faced a severe housing shortage, particularly for black workers and their families (Bowley, 1978).

2. The Federal Housing Authority (FHA) initially owned the development, and it was managed by the Chicago Housing Authority (CHA). The CHA did not become owner of the property until 1956 (Bowley, 1978).

3. Although this incident happened many years ago, very little has changed in Armour Square. In the summer of 1997, three young African American boys happened to ride their bicycles from nearby Stateway Gardens (CHA) to Armour Square Park. They were attacked by three white youths. Although two of the boys were able to get away, the third was severely beaten and suffered brain damage from which he may never recover.

4. T. E. Brown Apartments was owned and managed by the Progressive Baptist Church, the only religious institution in South Armour Square.

5. The "neighborhoods" of South Armour Square seemed to fit very closely with the face-block street neighborhoods identified by Jane Jacobs (1961). For example, residents who lived on Princeton Avenue would refer to their neighborhood as the Princeton neighborhood and talk about the people who lived only a block away as residents of the Wells neighborhood.

6. In 1971, as the result of tenant activism, the CHA entered into an agreement known as the Memorandum of Accord, which created a local advisory council (LAC) at each CHA development. The LACs are the officially recognized voices of residents and provide advice to the CHA about the development's budget and rehabilitation priorities.

7. Veeck, it was said, had a good relationship with the South Armour Square community. Ben Joravsky (1987) quotes Gus Zimmerman, a resident who first moved to South Armour in 1951, as saying of Veeck,

> Now there was a good man. [He] walked through this neighborhood all the time. He'd walk right up to your house and say, "Come on out and see my fireworks." Yes sir, Bill Veeck was some man. (p. 16)

8. According to Timothy Romani, who would become the deputy director of the Illinois Sports Facility Authority, Addison was the actual preferred choice of the White Sox, which led the city to feel that they would have to offer a very "sweet" deal to keep the Sox in the city. Whatever the actual intentions of the Sox were, they effectively used the threat of leaving the city for the suburbs or Florida as a tactic to get exactly what they wanted from both the city and state of Illinois (Gunner, 1991).

9. The Chicago Area Project was formed in 1932 in three of Chicago's highest crime areas to test delinquency prevention techniques. Today, it operates through more than 40 affiliates and special projects involving community organizing, educational support and counseling, recreation and cultural programs, leadership development, and advocacy.

10. City officials would later contend that no such promise was ever made (Joravsky, 1987).

11. When Comiskey Park was finally built, it became the last of the "modern" steel and concrete stadiums that proliferated throughout major league baseball beginning in the 1960s. It has features such as a steep upper deck and seats located far from the playing

field that have made it unpopular among fans. Subsequent new stadiums, such as Baltimore's Camden Yards and Cleveland's Jacobs Field, have been designed to "fit in" to the neighborhood and to recapture the spirit of "old-time" baseball. In retrospect, the design by Bass, which was scoffed at by both city officials and the White Sox, would have likely been better received by fans and baseball purists than the park that was eventually built.

12. Inviting Choca to participate in the conference was a tactical decision. South Armour members were angry with the city, but they wanted to show the city that they were willing to work with it if conditions could be changed. Choca was extremely nervous throughout her presentation, but conference attendees treated her with respect and applauded politely when she finished.

13. One reason alleged as to why the White Sox owners were so adamant about locating the park south of 35th Street was that it had been their intention all along to get rid of the Armour Square neighborhood. The Sox were having difficulty getting suburban residents to come to games, and one reason given for this was that suburbanites were fearful of what they perceived as a black slum so close to the stadium. Thus, it was argued, that by eliminating the "slum" and building a brightly lit parking lot on its former site, suburbanites might be more likely to come to Sox games.

14. By this time, Sheila Radford-Hill was no longer working for the Chicago Area Project, and her ability to devote significant time to the Wentworth residents was limited. She expressed her continuing commitment to the residents by organizing and delivering portions of the empowerment training workshops.

15. Success in achieving resident management came about as the CHA restructured as much as possible to get out of the business of managing public housing. Developments in Chicago that are not to be torn down have been given the option of either managing the units themselves or having a private management firm under contract to the CHA.

16. The lease agreement for the stadium gives the White Sox what many feel is a generous deal. The cost of building the new stadium was $119.4 million, and because the Sports Facilities Authority is a state agency, the city cannot collect the $488,000 annual property tax that the Sox would have paid. In addition, the Sox pay rent for use of the stadium only in years for which their total attendance exceeds 1.2 million. That means, when the team does not play well, it is likely that the Sox will have free use of the stadium (Wiltz, 1990).

# 7

# RESIDENT MANAGEMENT AND COMMUNITY EMPOWERMENT

The notion that residents living in public housing could successfully take control and manage the developments in which they live was especially popular in the 1980s and early 1990s, first being touted during the second Reagan administration and then heavily promoted by Jack Kemp, secretary of the Department of Housing and Urban Development (HUD) during the Bush administration. Proponents of resident management argue that it leads to improved housing quality, more efficient and cost-effective housing management, and tenant empowerment.

The beginnings of resident management of public housing can be traced to the early 1970s, when it arose almost simultaneously at public housing in Boston and St. Louis. In Boston, residents at the 1,100-unit Bromley-Heath development became managers after a long process during which they had assumed responsibility for the delivery of health and social services to residents and had developed a drug treatment center. Assuming control of the management of the development was just another step in an overall process of resident control. In St. Louis, five separate developments became resident managed between 1973 and 1975, after a 1969 rent strike resulted in a massive reorganization of the St. Louis Housing Authority.

Between 1976 and 1979, the Ford Foundation and HUD jointly sponsored a National Tenant Management Demonstration Program involving seven public housing sites in six cities. The national program was terminated when an evaluation found costs to be high relative to the benefits realized.

When resident management reemerged in the 1980s, it was tied to the notion of resident "empowerment." Federal legislation in 1987 and 1990 established procedures for creating resident management corporations (RMCs) and funding mechanisms for converting public housing developments from conventional housing authority management to resident management. Conservative proponents of resident management pushed the additional notion that resident management was a step on the way to tenants becoming owners of the developments. To them, homeownership was the ultimate "empowering" activity.

This chapter recounts the story of the struggle to obtain resident management at Leclaire Courts, a public housing development on the southwest side of Chicago. In telling the story, we see that the process of taking control of a public housing development is complex, multifaceted, and achieved slowly.

## WORKING TOWARD COMMUNITY: A BEGINNING

The Leclaire Courts public housing development is located on the far southwest side of Chicago about a mile north of Midway Airport (see Figure 7.1). The site, which had originally been vacant land adjacent to industrial uses, is far removed from other predominantly black communities of Chicago, but it is also isolated from surrounding white residential areas. Leclaire is one of eight Chicago Housing Authority (CHA) "court" projects that were financed by the city of Chicago and the state of Illinois in the early 1950s to meet the need for relocation housing resulting from urban renewal (Bowley, 1978).

The original construction at Leclaire consisted of 316 units in several two-story town houses with flat roofs. The site is relatively large, 24

# Leclaire Courts and Vicinity

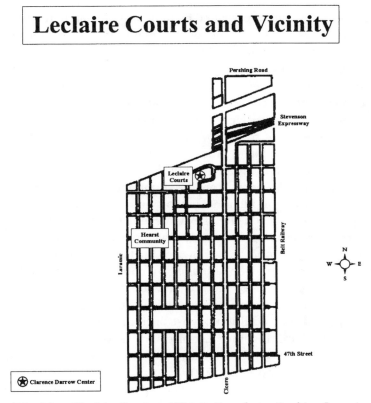

**Figure 7.1.** Map of Leclaire Courts and Vicinity Boundaries: Pershing, Laramie, 47th Street, and the Belt Railway

acres, and has suburban-like winding streets. Thus, the development is fairly attractive, despite the fact that the town houses look a lot like shoeboxes (Bowley, 1978).

Leclaire was expanded in 1954 when a federally financed 300-unit "extension" was constructed just south of the original development. The extension also consists of two-story town houses but with somewhat more attractive conventional pitched roofs. However, the extension is more dense and lacks winding streets, so that the net effect is to give the entire development an institutional feeling (Bowley, 1978).[1]

According to many longtime residents, Leclaire was well liked by its early occupants. Though the units and rooms are small, the relatively spacious setting was appreciated by families who had heretofore been living in densely packed and often deteriorated slum housing. At Leclaire, there was room to move about, and each family had a yard. Each household had the responsibility of maintaining the spaces in front of and behind their units, and many residents took the opportunity to plant attractive flower and vegetable gardens. In addition, plots for "serious" gardeners were created on several acres of open land at the northern edge of the property. Early residents knew each other, looked after each other's children, and participated in community activities found in any urban neighborhood.

Leclaire's location, away from the slums and close to job opportunities, was initially so popular that when many families no longer needed the support of public housing, they were reluctant to move away. Many chose to purchase homes in a newly developed area immediately to the west of Leclaire. This became known as the Hearst community, named after the local public school attended by both the children of Leclaire residents and by those of the homeowners. The Hearst community, consisting of modest yet well-maintained post–World War II brick bungalows, provided an additional sense of civility and neighborhood for families at Leclaire. Together, Leclaire and Hearst make up a community of just over 5,000 residents, nearly all of whom are African Americans (see Table 7.1 and Figure 7.2).

Despite its location, Leclaire was not spared the problems of public housing in Chicago that began to occur during the 1960s and 1970s. By the late 1970s, residents were complaining about leaking roofs, backed up sewers, and overall lack of general maintenance. Cutbacks in CHA personnel at the site resulted in delayed responses to tenant complaints. But perhaps even more important, residents perceived that the CHA had become lax in its selection and assignment of new tenants to Leclaire. "Problem tenants," it appeared, were no longer being screened out, and the CHA was no longer making home inspections prior to accepting new families. This seemed to be leading to rises in crime, drug abuse, prostitution, and gang activity throughout the development.

Around 1980, a group of residents, headed by a well-respected male tenant who had lived at Leclaire for years, began meeting and attempting to address the growing problems. Believing that nothing would be

**TABLE 7.1** Demographics of the Leclaire Courts Neighborhood

|  | Census Tract | | | |
|---|---|---|---|---|
|  | *5601* | *5602* | *5603* | *5604* |
| **1980** | | | | |
| Total population | 985 | 5,180 | 3,151 | 1,312 |
| % black | 5 | 97 | 0 | 0 |
| % of Spanish origin | 21 | 2 | 9 | 12 |
| % 13 years old and younger | 15 | 32 | 16 | 17 |
| % 65 years old and older | 11 | 2 | 11 | 8 |
| Median family income, 1979 | $22,614 | $15,437 | $27,109 | $26,414 |
| % income below poverty level | 8 | 32 | 4 | 4 |
| % white-collar workers | 61 | 44 | 49 | 43 |
| Population per household | 3 | 4.4 | 3.2 | 3.4 |
| Total housing units | 339 | 1,159 | 954 | 412 |
| % condominiums | 0 | 0 | 0 | 0 |
| % built 1970 or later | 4 | 3 | 0 | 0 |
| % owner occupied | 0.67 | 42 | 94 | 99 |
| Median value: Owner units | $45,200 | $39,400 | $48,300 | $48,400 |
| Median rent: Rental units | $236 | $140 | $288 | 0.00 |
| **1990** | | | | |
| Total population | 1,049 | 4,283 | 2,923 | 1,127 |
| % black | 0 | 100 | 1 | 0 |
| % of Spanish origin | 42 | 1 | 17 | 16 |
| % 13 years old and younger | 23 | 28 | 16 | 16 |
| % 65 years old and older | 11 | 5 | 18 | 17 |
| Median family income, 1989 | $31,172 | $17,360 | $42,107 | $42,386 |
| % income below poverty level | 11 | 35 | 3 | 11 |
| % white-collar workers | 37 | 49 | 50 | 55 |
| Population per household | 3.3 | 3.7 | 2.9 | 3.2 |
| Total housing units | 382 | 1,170 | 959 | 406 |
| % condominiums | 0 | 0 | 0 | 0 |
| % built 1980 or later | 1 | 1 | 0 | 2 |
| % owner occupied | 76 | 46 | 96 | 97 |
| Median value: Owner units | $64,400 | $63,500 | $75,300 | $71,400 |
| Median rent: Rental units | $470 | $233 | $629 | $604 |

NOTE: Data in this table are from the U.S. census, as organized and reported in the *Local Community Fact Book, Chicago Metropolitan Area 1990* (Chicago Fact Book Consortium, 1995).

## Leclaire Courts and Vicinity

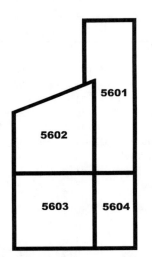

**Figure 7.2.** Census Tract Map of Leclaire Courts and Vicinity Tracts 5601, 5602, 5603, and 5604

solved as long as the CHA was the landlord at Leclaire, this group had a vision of taking over the development and converting it into cooperative apartments. However, the group lacked both the knowledge and expertise to proceed. And unfortunately, before much progress was made, the leader died and, temporarily, at least, so did the vision.

Dissatisfaction with conditions at Leclaire remained, and periodically attempts were made to seek improvements at the development. Residents, however, often disagreed over what should be done, and this resulted in the formation of at least three different "community groups," each claiming that they represented all residents.

Conditions at Leclaire continued to worsen, and in 1983 several female residents, who were leaders of the differing community factions, sought help from Stanley Horn, the director of the Clarence Darrow Community Center, a social service agency located at Leclaire and affiliated with the Hull House Association. As a condition of agreeing to provide help, Horn insisted that the women form a single group or organization whose purpose would be to work toward the improvement of the development. This laid the groundwork for what was later to become an RMC.

With Horn's assistance, the women prepared a list of goals they would like to see accomplished. One of these was to return to the earlier vision and to determine whether it would be feasible to convert Leclaire into a resident-owned housing cooperative. Recognizing that he lacked the knowledge of how to assess the feasibility of a cooperative conversion, Horn contacted a friend at the University of Illinois at Chicago, who in turn called me. At this point, the Voorhees Center and I began working with the residents on a project that would eventually result in their taking control of the development.

I assigned a graduate student to the project of preparing a feasibility analysis, and the director and staff of the Darrow Center continued to work closely with a small group of about a dozen interested female tenants. After some coaching, the newly formed resident group was able to hold a meeting attended by about 70 other residents and the regional director of the CHA who was responsible for overseeing the management of Leclaire. At this meeting, residents got the regional director to agree to pay more attention to management and maintenance problems at Leclaire. Although this led to only a few minor improvements at the development, it did show the residents that by working together, they were able to influence and affect officials at the CHA.

The result of the feasibility study was not encouraging. Because interest rates in 1983 were quite high, the computations showed that even if the CHA were to give Leclaire to the resident group, the costs of financing the rehabilitation work that was needed would result in monthly cooperative charges (rent) above $550 for a one-bedroom apartment. Such charges were clearly too high for public housing residents and were even higher than comparable market rate rents. Converting Leclaire to a cooperative, our study concluded, was impractical unless a sizable grant to underwrite rehabilitation could be obtained or continuing and deep rent subsidies could be ensured.

By this time, I had assigned a new graduate student to the Leclaire project, and he very diligently began looking for an alternative to cooperative conversion. Almost by chance he discovered the notion of resident management, then called tenant management.[2] He learned that tenant management had been tried in the 1970s, first in Boston, next in St. Louis, and then in several other cities as part of the federally sponsored National Tenant Management Demonstration Project.

Although the demonstration project had not been judged especially successful (Manpower Demonstration Research Corporation, 1981), and support for tenant management had somewhat subsided, there were signs that it might again become a popular idea. Apparent new successes with tenant management at the Cochran Gardens development in St. Louis and the Kenilworth Parkside development in Washington, D.C. were giving new life to the tenant management concept.

Beginning in 1984, we undertook several "field trips" to give the then loosely organized group of residents an idea of different types of community-based low-income housing possibilities that they might want to consider implementing at Leclaire. The first few trips were half-day excursions to sites in Chicago. Although the Leclaire residents were impressed with the quality of the housing they saw, they did not feel that the community-based housing we were showing them had much relevance to their situation as public housing residents.

But then we took a 2-day excursion to St. Louis to tour Cochran Gardens and meet with its dynamic resident leader, Bertha Gilkey. The 25 or so residents who went to St. Louis were greatly impressed with both the quality of the rehabilitation work at Cochran Gardens and the positive and enthusiastic outlook of Gilkey and other residents they met. The Leclaire residents immediately began saying how nice it would be if they could have resident management, but they also expressed doubt as to their own abilities to accomplish it.

It took one more trip to convince the residents that they could come together as a community to work for resident management at Leclaire. This time, they visited Kenilworth Parkside in Washington, D.C. On the first day of this visit, the Leclaire group was challenged by Kimi Gray, the resident leader at Kenilworth Parkside, to prove to her that they really wanted to make changes at Leclaire.[3] She even refused to talk with them unless they demonstrated to her that they were serious about resident management and were organized to accomplish it. Left alone in a room and after heated discussion, the residents agreed that they should form an organization, and one of the women, Irene Johnson, agreed, albeit reluctantly, to be the leader. What the residents realized as a result of Gray's challenge was that although they might get help from groups such as the Darrow and Voorhees Centers, the burden for convincing other residents of the merits of resident management and convincing the CHA that it was possible at Leclaire would fall on them. Thus, 800 miles from home, the Leclaire Courts Resident Management Corporation was born.

## WORKING TOWARD AND OBTAINING RESIDENT MANAGEMENT

Following the Washington trip, the Leclaire residents started a process that would take much work and nearly 3 years to complete. Efforts began in three different directions. First, the core group of women began working to convince their neighbors that resident management was desirable and achievable. Second, some 20 or so residents entered into a formal training program that would equip them with the knowledge and skills needed to create and sustain a viable not-for-profit housing management organization.[4] And third, the core group began a campaign to convince the CHA that the residents at Leclaire should be given a chance to become managers.

To convince residents who had not been involved in any of the work up to that time of the merits of resident management, the core group started a door-to-door campaign in which they attempted to reach every Leclaire household and explain to them the basics of resident management and how it could be accomplished. Household members were asked to sign a petition stating that they were aware of resident management and were supportive of the work being done to bring it to Leclaire. In time, at least one member of a majority of the households signed the petition.

Community meetings were held at least once a month to keep interested residents apprised of the progress being made. More than 200 residents attended a meeting at which Bertha Gilkey encouraged residents to push forward in their efforts. Like the meeting with Kimi Gray in Washington, the meeting with Gilkey was an important milestone and served to convince many skeptical residents that resident management was attainable.

Providing the core group with the knowledge needed to manage the development was another key element in the campaign to achieve resident management. The core resident group consisted of middle-aged African American women, none of whom had ever taken on an effort as large as managing a 614-unit housing development. These women were naturally concerned that their prior educational experience (most were high school graduates with no post–high school education) and previous community work had not prepared them for something that seemed to require professional training or a college education. In addition to their personal self-doubts as to their capabilities, they had to face the doubts of many other residents who felt nothing could be done to improve Leclaire and that any tenant efforts were destined to fail. The core group also got no support from the CHA, whose administrators were opposed to any attempts at self-management.

Initially, it was hoped that we could develop and provide the core group with a classroom curriculum for management training using resource specialists available in Chicago. When we approached the Amoco Foundation for funding to accomplish this, we were informed that they were already funding the National Center for Neighborhood Enterprise (NCNE) to do technical assistance with aspiring resident management groups. We thus had to settle for an NCNE curriculum and a trainer who traveled to Chicago once a week. We were soon to find that the NCNE was not really equipped to do management training, and thus the forced reliance on the NCNE resulted in the core group of residents being insufficiently prepared to assume the responsibilities associated with a board of directors of the resident corporation. The lack of training would become one of the problems that would ultimately lead to the failure of the effort.

We also formed a local advisory committee to provide assistance to the resident group. The advisory committee included specialists in a variety of areas of needed expertise, including housing management, legal issues, community organizing, and government operations. The individuals who agreed to be advisers provided countless hours of service, nearly all of them at no charge. The importance of these voluntary contributions to the success of the resident management effort cannot be overestimated. Not only did the residents feel more confident with a sophisticated group of advisers supporting them, but others, including CHA administrators, could not easily dismiss the resident efforts as merely a misguided attempt by a group of poor ignorant tenants.

Convincing the CHA to sanction resident management at Leclaire was probably the most difficult task facing the residents, their leaders, and their advisers. At the time, almost no one at the CHA knew anything about resident management, and most were skeptical of its practicality when told about it. Many CHA staff members refused to believe that residents could be capable of assuming management responsibilities, and some were suspicious of the residents' advisers, suspecting that they were merely using the residents to promote their own agendas.[5]

At the time, the CHA was preoccupied with its own serious internal problems. Charges of mismanagement and corruption, serious internal conflicts, and the apparent inability to both maintain and provide services to the developments had brought the CHA to the brink of collapse. Between 1983 and 1988, the executive directorship of the CHA changed five times. Each time there was a change, the Leclaire residents were forced to start almost from scratch to educate the new director and his or her staff about what resident management was, what the Leclaire Courts Resident Management Corporation (LCRMC) was, and why the LCRMC would soon be ready to assume management of Leclaire. By the spring of 1987, when the residents were prepared to make their bid to become resident managers, the situation at the CHA had deteriorated to the point where HUD was seriously considering placing the authority into receivership.[6]

The Leclaire residents did not let the problems of the CHA deter them from proceeding toward their resident management goals. They were forced, however, to educate each new director and his or her advisers about resident management, to continually address the opposition of many CHA staff, and to counter assertions that management was too difficult a task for "tenants" to handle. This was accomplished through an increasing sophistication of the resident leaders, resulting from the weekly training sessions and ongoing consultations with the growing cadre of advisers.

During informal and later formal negotiations, someone from the CHA would often raise an objection regarding some aspect of resident management. The residents, however, were nearly always ready to explain in detail why the objection was invalid. For example, at one point during the formal negotiation of the resident management contract, CHA lawyers objected to a demand that the RMC receive a management fee that would allow the RMC to independently make improvements to the development. The lawyers claimed a management fee made no sense because the residents were not a management company. Quick research found that not only did existing RMCs in other cities receive management fees but that in all instances, the fees they received were higher than those being requested by the Leclaire residents. When Irene Johnson, Leclaire's leader, presented this information at a subsequent negotiating session, the CHA was forced to yield the point without further argument.

The CHA often proved untrustworthy during the long negotiating process. For example, at a May 1986 board meeting, CHA directors passed a resolution that purported to give the residents of Leclaire control over maintenance, security, and tenant selection and to set aside $1 million of capital funds for property rehabilitation at Leclaire. In a symbolic gesture, which was reported in all of Chicago's daily papers, the residents were presented with an oversized check for the $1 million. After months of trying to get the CHA to make arrangements to transfer the $1 million to an account controlled by the residents, CHA officials admitted that there had never been any money and that the board action and presentation of the check had been little more than a publicity "gimmick." For several years, the residents kept the "bad check" prominently displayed on the wall of the RMC office as a reminder of their constant need to be dubious about the claims and actions of the CHA.

Over time, most of the objections raised by the CHA and arguments as to why resident management would not work were countered by the residents. After exhausting all of its options, the CHA was finally forced to begin seriously negotiating toward a contract with the LCRMC that would give control of Leclaire to its residents.

After numerous broken promises and delays, on September 23, 1987, the CHA board of directors formally approved a contract establishing a 1-year transition period of dual management between the CHA and the LCRMC, to be followed by full resident management. At the signing of the contract a few days later, resident hopes were high. What had started as an idea and a hope on the part of a few unsure African American women had become a community endeavor, supported by an overwhelming majority of the residents at Leclaire. Leclaire, it seemed, was again becoming a good place to call home.

As the RMC took over the management, it could point to several already obtained or soon-to-occur improvements at Leclaire. Roofs, many of which had leaked badly, were repaired, and serious plumbing problems were attended to. Through a grant that the RMC received from the state of Illinois, new energy-efficient windows were placed in all units. Working with the Clarence Darrow Center, on-site health care facilities were improved. Resident committees were formed to address problems of community safety, economic development, and education, and new programs such as a bus service to transport residents to suburban jobs were initiated. Clearly, resident management meant more than just a few residents receiving jobs. It promised to bring about a whole new and better future for the community.

## LECLAIRE IN A BROADER CONTEXT

Ford (1991) suggests that changes occurring in urban areas can only be fully understood by considering actions at a variety of levels, which he classifies into three categories: macro, meso, and micro. Resident management at Leclaire Courts did not happen simply because a group of longtime residents became dissatisfied with conditions and trends at the development and banded together to do something about it. True, they were central to achieving control of the development. However, achieving success was a more complex process and should be thought of in the broader context of events and actions involving the residents, their advisers, the CHA, and for that matter federal policy regarding public housing.

By the 1980s, public housing in Chicago truly had come to be what Bowley (1978) called the *poorhouse*. This sorry state resulted from direct and indirect actions by federal, state, and local officials responding to calls from Chicago's civic leaders to help shore up the city's central core through urban renewal and from leaders in white neighborhoods who wanted to maintain racial segregation (Hirsch, 1983). This resulted in the concentration of public housing in existing slum areas or on isolated land that no one else wanted. Once housed in poorly designed and inadequately maintained buildings and out of sight, public housing was allowed to deteriorate both physically and socially. The residents of Leclaire did not make a conscious choice to live in a community that was isolated, poor, and a locus of economic and social problems. The historical events and the actions of many individuals, organizations, and government agencies largely created the place known as Leclaire and gave it its character.

Likewise, it was not possible for the residents acting alone to change conditions at Leclaire. Even though they were central and indispensable to the resident management effort, they would not have succeeded without the assistance and support of others, especially their advisers and even, when necessary, the CHA. At the time the residents were working to take over management, many people, from federal policymakers to average citizens, were aware that public housing had problems and were supportive of ideas and programs that would address public housing needs. Thus, when residents at Leclaire discovered resident management, they did so in a climate that was more or less receptive to both their plight and the notion that something should be done. This, at least in part, allowed them to advance resident management at Leclaire from an idea to a reality. Even the initial reluctance of the CHA to give serious consideration to the residents' wishes acted to spur them on to become more knowledgeable and to ultimately prove that they could manage the development.

The efforts at Leclaire should also be seen in the context of a larger resident management movement that existed at the national level during the 1980s. Although resident or tenant management had been tried before in the 1970s, it struck a responsive chord with conservative policymakers and politicians, who saw it as a way of promoting both self-help and as a solution to troublesome public housing.[7] Likewise, it was popular with liberals who saw resident management as a way to involve residents in their own governance. Thus, to nearly everyone, it appeared to be a solution to the problem of what to do with public housing.

Promoted by conservative "think tanks" such as the National Center for Neighborhood Enterprise, the resident management movement rapidly gained steam during the second Reagan administration. But it was during the Bush presidency that it received its highest visibility. Resident management as championed by HUD's persuasive secretary, Jack Kemp, became almost the only public housing initiative during at least the first 2 years of the Bush administration, and residents of public housing developments throughout the country began to train to take control over management. A study commissioned by HUD near the end of the Bush administration (ICF, Inc., 1992) identified 11 developments as having "mature RMCs" and being fully managed by a resident corporation, as well as an additional 27 "emerging" resident management corporations that had some degree of management control over their developments. In Chicago, the experiences at Leclaire stimulated residents in other CHA developments to explore the possibility of resident management, and resident corporations were formed at approximately one third of CHA developments.

Once again, more than a decade after Leclaire became resident managed, resident management no longer seems to be a popular solution to the problems of public housing. The Clinton administration never embraced resident management with as much fervor as did the Bush/Kemp regime, and current public housing policy is more focused on the restructuring of HUD and the recasting of public housing into developments with fewer units, targeted, where possible, for mixed-income settings. In such situations, resident management seems either inappropriate or irrelevant. The Clinton administration seems as determined to get the federal government out of the public housing business as were its Republican predecessors but has apparently chosen a different mechanism to achieve the goal.

Resident management at Leclaire Courts, after enjoying early popularity and success, stumbled, entered into a period of slow decline, and ultimately failed. In 1995, the CHA, citing irregularities in the RMC's books, recaptured the management at Leclaire by dissolving the RMC and reinstating its own management team.[8] How did something that was initially so successful and promising ultimately fail?

During the first year of resident management at Leclaire, the period of transition between CHA and resident managers, a new and reform-minded individual, Vincent Lane, became executive director and chairman of the board of the CHA. Among Lane's qualifications were that he had been involved in a study of resident management conducted by the influential Chicago civic group, the Metropolitan Planning Council, and he had also been a member of Chicago Mayor Harold Washington's advisory council on the CHA. He was knowledgeable about resident management and supported it. One of his first actions as director was to create a new department within the CHA to assist the several resident management initiatives that were springing up throughout CHA developments.

Although Lane's arrival and support of resident management would seem to be a positive occurrence, in reality it had a negative impact on Leclaire. Because the CHA now offered technical support for resident management efforts, there no longer seemed to be a need for the many advisers that had been assembled to assist the Leclaire RMC. In fact, CHA staff and HUD officials actually discouraged the use of outside advisers, arguing that the CHA could and would provide all of the assistance the RMC needed.

One by one, the RMC's advisers fell away, either because they were ignored or told outright that they were no longer needed. First the RMC dismissed an experienced community organizer who had been hired to further strengthen the community's support for resident management. Next it chose to ignore a plan for structuring management activities that had been prepared by a real estate specialist with many years of experience of managing cooperative housing. No further assistance was sought from the Voorhees Center. Finally, Stanley Horn, the direc-tor of the Clarence Darrow Center who had been so instrumental in developing the RMC, resigned because the director and board of the RMC began to reject or ignore his advice.

The RMC no longer had to prove to the CHA that it could manage Leclaire, and it became complacent. Leclaire had become the CHA's model of success. Interestingly, however, the CHA made several attempts to weaken the authority of the RMC by attempting to renego-tiate or ignoring parts of the management contract.

Another weakness arose when the RMC's resident leader, Irene Johnson, began to distance herself from the board of directors. The training of the board had never been completed, and the board did not feel secure in making key decisions. Johnson was frequently called to Washington to meet with Jack Kemp and occasionally with the presi-dent. As she became more and more associated with resident manage-ment as a national "movement," she began to make more unilateral decisions about Leclaire without consulting the board. These actions resulted in an estrangement between the board and its director. To the board members, Mrs. Johnson was their friend and neighbor, and they were reluctant to criticize her. But factions were developing within the board and among other residents who supported one side or the other.

The quality of resident management at Leclaire slowly declined. Some of the RMC's early initiatives in the areas of health care and economic development were terminated. Record keeping deteriorated, and by the time the CHA stepped in and took back the management, the RMC was considerably behind in meeting its financial obligations.

That resident management at Leclaire may not have fully lived up to its initial expectations and that it ultimately failed should not suggest that the efforts of the residents were unimportant or in vain. Leclaire residents did have a period of nearly 10 years during which they had greater control over what happened in their development than they had in the past. Numerous physical improvements, including sewer repairs, new roofs, new windows, and interior remodeling of many units, can be directly attributed to the RMC effort. Leclaire residents had shown that by coming together, they could create or re-create a community. That they succeeded, if only for a while, suggests that it is possible, even against considerable odds, for poor people to work together to realize their dreams for their community. Even though Leclaire is no longer resident managed, it is a better place than it was before because good people chose to do good things, and others recognized this and offered their assistance.

*   *   *

The struggle to build a community through resident management at Leclaire Courts demonstrates the effectiveness of community development planning when it is accompanied by community organizing and sufficient internal and external resources. Although the residents at Leclaire Courts were the core of the community development effort, they would not have succeeded without the assistance of many dedicated individuals who provided technical and political assistance and without the receipt of grant moneys sufficient to purchase what was needed to do battle with the CHA. In the end, it was a team effort, and Leclaire remained viable as long as a team mentality existed. When the residents began to isolate themselves and to rely too strongly on the resources of the CHA, their efforts faltered.

## NOTES

1. Technically, Leclaire Courts is two separate developments, the "city-state" and "extension" sites. Residents, however, view it as a single site, and the resident organizing effort made no distinctions between the two. The fact that the two developments have different sources of financing and maintain separate sets of financial records created a minor, though not insurmountable, obstacle to the resident management effort.

2. The search for an alternative form of management was supported by a modest, yet crucial, grant from the Amoco Foundation.

3. I was not present during this trip but have frequently heard this story recounted by those who were there, both from residents and Darrow Center staff. Although it does have the ring of being apocryphal, it is nonetheless clear that whatever actually occurred, this meeting with Kimi Gray was the turning point in convincing residents that they wanted to work toward resident management at Leclaire.

4. This group would become the board of directors of the incorporated Leclaire Courts Resident Management Corporation.

5. To be fair, it should be noted that not all Chicago Housing Authority (CHA) personnel were negative about resident management. Local site managers, who knew both the residents and the conditions of the developments, were more likely to be sympathetic and supportive of resident efforts than management staff from the "central" office. The site manager at Leclaire became an early supporter of resident management and ultimately became an adviser to the resident group. When the Leclaire Courts Resident Management Corporation (LCRMC) assumed management responsibilities, she took a yearlong leave of absence from the CHA and became a paid consultant for the LCRMC. She was an invaluable resource during the first year of resident control.

6. Takeover by Department of Housing and Urban Development (HUD) was forestalled by Chicago's mayor Harold Washington, who appointed a "blue ribbon" advisory council on the CHA in early 1987 to "assist the CHA to improve its operations, to bring about improved services to Authority residents, and to expand the availability of quality low-income housing for eligible individuals" (Advisory Council on the Chicago Housing Authority, 1988, p. 1). It was only the timely intervention by the cochairs of this committee who traveled to Washington to meet directly with top HUD officials that staved off receivership at that time. HUD would later, as conditions at the CHA again deteriorated in the mid-1990s, be forced to assume management control of the CHA.

7. As a policy issue, resident management can be seen as part of a conservative effort to move the federal government out of the business of providing housing. For conservatives, resident management was seen as a stage in a process that moved residents from welfare-dependent renters to independent homeowners. Not incidentally, as residents became owners, the inventory of public housing would be reduced.

8. Technically, Leclaire Courts is still resident managed. In reclaiming Leclaire, the CHA dissolved the RMC's board of directors and fired its chair and president of the corporation, Irene Johnson. A new board, with much reduced powers, was subsequently selected. The new board acts as an adviser to management, rather than the body with full control over management decisions and actions.

# 8

# REBUILDING ROSELAND A BLOCK AT A TIME

Roseland, one of Chicago's 77 community areas, is located approximately 11 miles directly south of the Loop, the city's central business district (see Figure 1.1). For the most part, it is a residential community, consisting of single-family homes and a few small apartment buildings. Some homes, such as those in the Rosemoor neighborhood at the northern end of Roseland, are one- and two-story brick buildings, well maintained and occupied by households with middle-class incomes.[1] Indeed, the median family income for Roseland, as measured in 1989 for the 1990 census, was approximately $31,000, which was just about the median for the city of Chicago and high for the mostly African American south side of the city. Homeownership, at 66% of all housing units, was also high, well above the 42% rate for the city as a whole (see Figure 8.1).

Despite its more prosperous areas, most people associate Roseland with poverty and distressed housing. This is certainly true about the southern and western parts of the community area, but even in this part

## Roseland

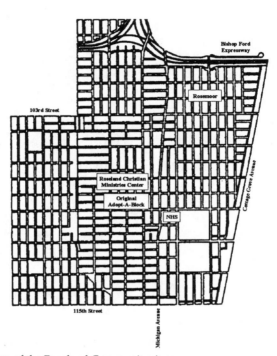

**Figure 8.1.** Map of the Roseland Community Area

of Roseland, home ownership rates are high. But gang activity and crime rates are also high. Here the houses, mostly of frame construction, are likely to be poorly maintained, and poverty and joblessness are rampant. For example, in the census tract in the southeast corner of Roseland (see Table 8.1 and Figure 8.2), the median family income in 1989 was only $18,600, and nearly 3 of 10 families were living below the poverty level.

Like many other low-income areas in Chicago, Roseland was not always so depressed. Settled in the middle of the 19th century by Dutch immigrants, the community started out as a pastoral village surrounded by farmland on which residents planted vegetable crops and flowers. Originally called Hope, the name was changed to Roseland

**TABLE 8.1** Demographics of the Roseland Community

|  | Census Tract | | | | | | | |
|---|---|---|---|---|---|---|---|---|
|  | 4907 | 4908 | 4909 | 4910 | 4911 | 4912 | 4913 | 4914 |
| **1980** | | | | | | | | |
| Total population | 4,226 | 5,503 | 12,661 | 8,972 | 6,857 | 3,388 | 4,721 | 4,816 |
| % black | 99 | 98 | 97 | 98 | 98 | 99 | 96 | 92 |
| % of Spanish origin | 0 | 0 | 2 | 1 | 1 | 0 | 1 | 2 |
| % 13 years old and younger | 23 | 21 | 24 | 30 | 24 | 26 | 30 | 29 |
| % 65 years old and older | 5 | 4 | 4 | 3 | 6 | 4 | 3 | 4 |
| Median family income, 1979 | $23,562 | $22,966 | $19,063 | $17,041 | $24,484 | $23,787 | $17,656 | $13,036 |
| % income below poverty level | 11 | 15 | 18 | 23 | 9 | 10 | 23 | 21 |
| % white-collar workers | 42 | 49 | 58 | 44 | 54 | 53 | 47 | 48 |
| Population per household | 3.7 | 3.2 | 3.2 | 4.1 | 4.1 | 4.3 | 4.1 | 3.2 |
| Total housing units | 1,078 | 1,590 | 4,174 | 2,163 | 1,613 | 831 | 1,244 | 1,621 |
| % condominiums | 0 | 0 | 0 | 0 | 0 | 0 | 0 | 0 |
| % built 1970 or later | 5 | 2 | 3 | 3 | 1 | 4 | 2 | 2 |
| % owner occupied | 0.88 | 80 | 55 | 69 | 0.88 | 83 | 56 | 30 |
| Median value: Owner units | $35,400 | $36,900 | $37,900 | $26,800 | $41,600 | $39,800 | $28,800 | $31,600 |
| Median rent: Rental units | $280 | $260 | $244 | $277 | $200 | $257 | $269 | $233 |

(Continued)

**TABLE 8.1** (*Continued*)

|  | Census Tract | | | | | | | |
|---|---|---|---|---|---|---|---|---|
|  | *4907* | *4908* | *4909* | *4910* | *4911* | *4912* | *4913* | *4914* |
| **1990** | | | | | | | | |
| Total population | 3,795 | 4,640 | 11,244 | 7,608 | 6,857 | 3,388 | 4,721 | 4,816 |
| % black | 100 | 96 | 99 | 98 | 98 | 99 | 96 | 92 |
| % of Spanish origin | 0 | 0 | 0 | 0 | 0 | 0 | 1 | 0 |
| % 13 years old and younger | 11 | 10 | 13 | 13 | 18 | 23 | 24 | 26 |
| % 65 years old and older | 9 | 11 | 7 | 6 | 6 | 4 | 3 | 4 |
| Median family income, 1989 | $36,071 | $34,087 | $33,131 | $24,234 | $42,944 | $42,261 | $22,700 | $18,636 |
| % income below poverty level | 10 | 10 | 14 | 24 | 9 | 10 | 23 | 21 |
| % white-collar workers | 48 | 61 | 60 | 59 | 58 | 61 | 50 | 47 |
| Population per household | 3.7 | 3.2 | 3.2 | 4.1 | 3.7 | 3.5 | 4 | 3.2 |
| Total housing units | 1,083 | 1,559 | 3,852 | 2,164 | 1,645 | 781 | 1,293 | 1,523 |
| % condominiums | 0 | 0 | 0 | 0 | 1 | 0 | 0 | 0 |
| % built 1980 or later | 0 | 0 | 0 | 0 | 1 | 0 | 2 | 0 |
| % owner occupied | 89 | 80 | 59 | 67 | 89 | 88 | 55 | 29 |
| Median value: Owner units | $56,100 | $63,900 | $63,300 | $46,100 | $65,100 | $61,900 | $45,500 | $51,900 |
| Median rent: Rental units | $521 | $432 | $463 | $480 | $610 | $555 | $536 | $456 |

NOTE: Data in this table are from the U.S. census, as organized and reported in the *Local Community Fact Book, Chicago Metropolitan Area 1990* (Chicago Fact Book Consortium, 1995).

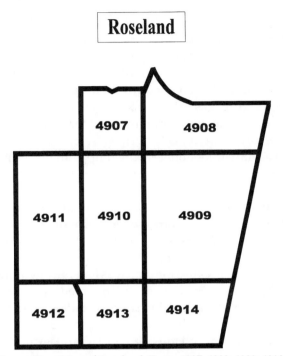

**Figure 8.2.**   Census Tract Map of Roseland: Tracts 4907, 4908, 4909, 4910, 4911, 4912, 4913, and 4914

because of annual displays by the villagers of brilliant red roses (Chi-cago Fact Book Consortium, 1995).

The city of Chicago soon grew out to meet the little town of Rose-land. George Pullman built his railroad car factory and company town just east of the community, and the Illinois Central Railroad con-structed its Burnside shops in the northeast corner of Roseland, on land now occupied by Chicago State University. Other industries followed, locating east, south, and west of the community. In 1889, Roseland was annexed into the city of Chicago.

Roseland continued to grow, increasing in population through the 1930s and 1940s.[2] Between 1940 and 1950, blacks began slowly moving into the community, at first occupying the more substantial housing in the northern portion of the community.[3] By 1960, blacks accounted for just over a fifth of the population of Roseland.

In the 1960s and the early 1970s, Roseland underwent a more dramatic change, caused to a considerable extent by a combination of mortgage redlining and Federal Housing Authority (FHA) mortgage fraud (see Bradford, 1979). Banks redlined Roseland, refusing to lend there, fearing that racial transition would result in community decline and defaulted mortgages. This discouraged white households from seeking to buy homes in the community and forced black home seekers to obtain federally insured FHA mortgages.

Some real estate professionals teamed up with unscrupulous mortgage bankers and, seeking to profit from racial turnover, sold houses to unqualified black families using FHA loans to finance the purchases. FHA officials were lax in checking applicant qualifications, which were often falsified. As the new homeowners were often unprepared and unable to make the necessary loan payments, they quickly defaulted on their loans. Although the real estate agent got his commission and the lender got his money because the loan was federally guaranteed, all the community got were abandoned and boarded-up homes resulting from the numerous defaulted loans. Even today, Roseland leads all other community areas in Chicago in the number of abandoned homes due to U.S. Department of Housing and Urban Development (HUD) foreclosures on bad FHA loans. In 1998, there were 169 properties owned by HUD as a result of FHA foreclosures, more than double the amount in any other community area in the city.

As lower-income blacks moved into the southern part of Roseland and as FHA fraud took its toll, the community began to decline. The main shopping area of Roseland, which runs north and south along Michigan Avenue through the community and was known as "the place" to shop on the far south side because of the location of Sears and Gately's Peoples[4] stores, rapidly deteriorated. Today, with a few exceptions, the Michigan Avenue shopping strip consists of vacant land (especially in the northern end of the street), empty buildings, a few small clothing stores, storefront churches, and taverns.

The economy of Roseland was affected by the closure of many of the industries that had been its job base. Factories such as the Pullman plant, Wisconsin Steel, Dutch Boy Paint, and International Harvester, all located near Roseland, have closed in the past 20 years, leaving local residents with far fewer job opportunities than had previously existed. Job losses, housing abandonment, deterioration of the business district, and general decline of the socioeconomic well-being of the residents have all worked to create a community desperately in need of renewal.

## THE CHRISTIAN REFORMED CHURCH AND ITS CHANGING PRESENCE IN ROSELAND

Historically, the Christian Reformed church, with its Dutch origins, has had a significant presence in Roseland. At one time, there were eight Christian Reformed churches located in the community. The building that housed the church's broadcast radio ministry, the Back to God Hour, and the Christian Reformed World Relief Commission was located at the corner of 108th Street and Michigan Avenue in the heart of Roseland.[5] This building was adjacent to the blocks that, during the 1960s and early 1970s, underwent the greatest amount of decline.

As racial transition, hastened by FHA mortgage abuse, swept through the core of Roseland in the 1960s and into the 1970s, white members of the eight churches fled westward out of Chicago and into the southwestern suburbs.[6] There was interest on the part of the Back to God Hour and Relief Commission staff to do likewise. In 1974, a request was made to the Church Synod, and permission was received to build new facilities in the southwest suburbs. The stage was set for yet another empty and deteriorating building along the once-thriving Michigan Avenue commercial strip.[7]

Some members of the Church Synod felt uncomfortable about their potential flight from Roseland, and church officials were instructed to find a "good Christian use" for the building, preferably some kind of a Christian Reformed ministry (Janke, 1996).[8] After some discussion, it was decided to establish a "ministry center" at the site. The resulting Roseland Christian Ministries Center, initially started as a 3-year pilot project in 1975, prospered and has grown to become a major social service resource in the Roseland community.

The Roseland Christian Ministries Center did not actually get off the ground until the summer of 1976 with the operation of a children's program. Soon thereafter, Reverend Anthony van Zanten was called to become the center's director. Van Zanten, who came with 11 years of experience with a similar program in Patterson, New Jersey, used the building's central location to initiate a mixture of religious and social service activities. As is typical, these early efforts were small and handled mostly by volunteers.

Over time, the center grew. I first learned about it in 1995, when I moved from the University of Illinois at Chicago to become the coordinator of Chicago State University's Neighborhood Assistance Center. By then, the Roseland Christian Ministries Center was operating a soup kitchen, providing hot meals for the homeless in the community, an overnight homeless shelter during winter months, a year-round drop-in center for needy men and women, a thrift store, a food pantry, a jobs counseling and referral service that included job training, a senior citizens club, and the coordination of the activities of trained volunteer teams that provided home repairs for neighborhood residents.

The center was only one part of a three-part program of the Christian Reformed church in Roseland. Reverend van Zanten had also reinvigorated the Roseland Christian Reformed church, which in addition to holding regular services had a boys and girls club, had a summer day camp, and was involved with several substance abuse programs such as Alcoholics Anonymous and Narcotics Anonymous.

The third part of the program, and the one that is the focus of this chapter, is the Roseland Christian Homes Corporation, a small community development corporation that initially set out to reclaim Roseland's decaying housing stock a home at a time.

## ROSELAND CHRISTIAN COMMUNITY HOMES CORPORATION: FROM ADOPT-A-HOME TO ADOPT-A-BLOCK

In 1979, at the urging of a seminary intern, the Roseland Christian Ministries Center created a task force to address the issue of housing abandonment in the community.[9] Initially, the task force tried to get the city to tear down houses that had been abandoned for so long that they were beyond repair. It was successful in having four houses torn down. Then it was given a house[10] by an owner who no longer wanted it, and the task force suddenly became involved in the business of housing redevelopment.

After remodeling this first house and selling it, there was money to buy another house and fix it up. This led to yet another house. Building on the early successes and wanting to do more to address the housing needs of the community, the task force restructured itself in 1988, becoming the Roseland Christian Community Homes Corporation (hereafter referred to as Roseland Homes), a nonprofit corporation with its own board and mission, which was to revitalize the housing stock of Roseland. Roseland Homes, though separate from the Roseland Christian Ministries Center, continued its affiliation with it and located its offices on the second floor of the ministry building.

To head the newly formed corporation, the board hired a Roseland resident, Maurice Williams, as its director. Williams had grown up just about a block from the center, was a longtime participant in its programs, and as a youth had become a member of the Christian Reformed church.[11]

The core of Roseland Homes program during its first decade of existence as a community development corporation was its "Adopt-a-Home" program, which allowed more than 30 low-income families to purchase single-family homes[12] that had been acquired and renovated by the corporation. Financing of the purchase and renovation of the homes was done somewhat differently than was typically done with most other Chicago-area community development corporations. Each Roseland Homes' home was "adopted" by a partner, with the partner commonly being a church. The Christian Reformed churches in Chicago's southern and southwestern suburbs were especially good sources of partnerships because most of them had member families who had once lived in Roseland before it had undergone racial transition. Reverend van Zanten and Williams successfully took advantage of this "link" to the old neighborhood to develop the partnerships.

The adopting partner provided $5,000 toward the down payment on the home, and the purchasing family was required to invest 200 hours of "sweat equity" in their home and an additional 100 hours of "sweat equity" in the next home purchased by the Roseland Christian Homes Corporation. Renovation costs were lowered further through the use of Roseland Homes' own "Revive" construction crew, young men from the neighborhood who were able to get employment and training opportunities through the corporation. The partnership arrangement, sweat work, and use of the Revive crew allowed Roseland Homes to offer the renovated housing for sale at a much reduced rate,[13] thus making homes available to low-income families in need of decent and affordable housing.[14]

But still, the family purchasing the home needed a mortgage to be able to pay their share of the home purchase. Here again Roseland Homes turned to its links with Christian Reformed churches and their members. The president of the Evergreen National Bank had been a longtime supporter of the Roseland Christian Ministries Center, and it was arranged that mortgages would be provided through his bank. Because Roseland Homes would be carefully screening the home purchasers, these loans were very likely to be sound, even if they were for amounts well below what the bank was used to lending. The partnership between the bank and Roseland Homes eventually developed to the point where the bank placed one of its loan officers at the center, thus making it even easier for potential homeowners to carry out the loan transaction.[15]

I met Maurice Williams in 1995 at a reception held at Chicago State University shortly after I was hired to coordinate the university's Neighborhood Assistance Center. A month or so later, he called me, saying that he wanted to discuss with me an idea for a new initiative. He called the new initiative the "Adopt-a-Block" program.

Although Williams was pleased that Roseland Homes had been able to make homeowners of 30 Roseland families and to provide them with decent and affordable housing, he was concerned that the Roseland Christian Homes Corporation was having little impact on the overall quality of housing and of life in general in the community. Although families in the Adopt-a-Home program got a remodeled house, the homes were often located on blocks where there were gang and drug activities, vacant and abandoned lots, and many other houses in need of repair. In addition, looking beyond the immediate blocks where Roseland Homes had been active, Roseland as a whole was still a depressed community and by some measures was getting worse.[16]

Williams's proposal was to expand the Adopt-a-Home program into an Adopt-a-Block program and to attempt to address the complete needs of a full block—physical, social, and economic. Although he recognized that fixing up just one block probably would not do that much more toward the revitalization of Roseland than fixing up a single house here and there, it would focus the Roseland Christian Homes

Corporation's efforts in a specific targeted area and would provide a better immediate environment for families moving into Adopt-a-Home structures. Over time, perhaps, as Roseland Homes expanded the program and began to revitalize block after block, there might also be observable community improvement.

As a result of our discussion, I agreed to become a member of an advisory committee that Williams was forming as part of the planning process for the Adopt-a-Block initiative. The committee consisted of local residents as well as individuals from outside of Roseland who had skills and expertise that might be of use in the development of Adopt-a-Block. I also agreed to identify a student at Chicago State, who would become a student intern assigned to Roseland Homes to specifically work on the planning for the Adopt-a-Block.

I quickly found a student, and she began her internship at the start of the new year, 1996. By then, the 109th and Wabash Avenue block, between Michigan Avenue and State Street, had been chosen as the target for the first Adopt-a-Block. This block was chosen because it was quite near to the Roseland Homes office in the ministries building at 108th and Michigan, there were existing working relationships between Roseland Homes and several families on the block,[17] the very first Roseland Homes Adopt-a-Home in Roseland was located on the block, and Roseland Homes was in the final stages of preparation for rehabilitating another Adopt-a-Home on the block. Thus, several of the elements believed critical to the success of the Adopt-a-Block initiative were already in place.

The Adopt-a-Block was intended as a comprehensive community development initiative for the entire block. The goals for the initiative, as developed early in the planning process, were to do the following:

1. rehabilitate all abandoned housing units on the block;
2. complete all interior and exterior home improvements on the block;
3. assist all unemployed residents of the block with job readiness skills and job placement services;
4. participate or assist in the coordination of a block club for the block;
5. provide spiritual development to any and all interested persons;
6. provide health care assistance to all interested individuals and families;
7. assist all interested persons on the block with education and training services (e.g., tutoring, mentoring, GED, etc.);

8. provide the residents with resources to prevent crime, drugs, and gang
   activity;
9. plant flowers, grass, trees, and shrubs throughout the block as signs of
   beauty;
10. build new housing on all vacant lots found on the block.

Taken as a whole, the successful achievement of these goals represented a serious challenge for the small development corporation. But
those of us involved in the Adopt-a-Block initiative felt that a program
of this magnitude was called for if the goal of Roseland Homes was to
pursue a meaningful strategy for community revitalization.

Planning for the Adopt-a-Block continued through the winter of
1996 with a date in April targeted for "kicking off" the program. From
the beginning, it was obvious that Roseland Homes could by itself not
be responsible for carrying out all elements of the program. Other
organizations and individuals, especially the residents of the target
block, would have to become involved. Nor would it be possible to
fully fund the program from the proceeds of Adopt-a-Home sales and
partnership gifts from other Christian Reformed churches, so new
sources of funding would need to be found.

So the planning process led to several different activities. Residents
on the block were contacted, and their involvement was solicited
through the formation of a block club. Other agencies and organizations in the community, including the local alderman and other civic
leaders, were contacted and endorsements sought. A funding plan was
also developed that targeted Chicago-area charitable foundations.

Early estimates indicated that it would take slightly more than $1
million to successfully revitalize the block. Funds would be needed to
rehabilitate 2 homes, make improvements in all 28 homes on the block,
construct 6 new homes, beautify 34 lots, and upgrade the block's infrastructure.[18]

But a holistic revitalization of the block would take more than
money. The block's residents would have to become involved and
would have to stay involved after the formal Adopt-a-Block activities
ended. One of the first things the Chicago State intern did was to
develop and administer, with the assistance of two student volunteers
from the university, a needs survey to all residents on the block. This
served not only to find out what resident concerns and wishes were but
also to alert the block members about the upcoming program. The sur-

vey was followed by an open meeting in March for all block residents seeking their input and involvement. Sixty-five individuals attended.[19]

During this planning time, the intern was also doing other necessary background work, such as identifying who owned the vacant lots and properties on the block, identifying and contacting other Roseland-area organizations that might be able to assist the Adopt-a-Block effort, and assisting in the planning for the April kickoff event.

The Adopt-a-Block was formally initiated at a daylong kickoff held on Saturday April 20, 1996. More than 150 volunteers were present, including residents, members of Roseland Christian Ministries Center, volunteers from Evergreen Bank, Chicago State University students, and students from Hope College in Holland, Michigan. The event was more than symbolic, as the volunteers did a general cleanup of the streets and alleys and painted the porches of 13 of the block's 28 homes. The intent was to show that Adopt-a-Block was more than a slogan and speeches and that it would result in immediate improvements for the block's residents.

According to the concept as initially designed, the Adopt-a-Block effort was to take 2 years to complete. Once a block was upgraded, the formal program for that block would end. The assumption was that an improved and organized block would be able to stand on its own and Roseland Homes would be able to move on and "adopt" other blocks. Eventually, as an increasing number of blocks became improved, there would be a spillover effect in the neighborhood, and blocks adjacent to Adopt-a-Block sites would begin to improve without direct intervention by Roseland Homes.

The Adopt-a-Block strategy was a big undertaking for an organization as small as Roseland Homes. Significant external resources would be needed to successfully revitalize even just the first block, let alone several blocks if the program were to be expanded. Funding was particularly needed for the new construction planned for the block. Therefore, much of the early effort was focused on proposal writing and fund-raising. At advisory board meetings, Williams would always report on the number of proposals submitted, the number outstanding, and the amount of money received. Although some money was raised, almost from the beginning it appeared that it would not be possible to fund the Adopt-a-Block at its anticipated full level.

But many of the aspects of the Adopt-a-Block strategy did not depend fully on external resources. Little or no external funds were needed to organize the block club, and after some initial work, the block club began meeting regularly. Assisted by the former student, Allegra Henderson, who had been hired by Roseland Homes after she had completed her internship and had graduated from Chicago State in the spring of 1996, the block club began putting out a bimonthly newsletter that contained a mixture of news about the block and helpful hints about community improvement. The early efforts by both Roseland Homes and the block's residents resulted in increased morale and cohesion among households on the block.[20]

Another Chicago State University intern was recruited to prepare a Roseland "resource guide" for the Adopt-a-Block. The guide contained information about a variety of services available to Roseland residents. It was the first document of its kind for the community. The guide was distributed to all families living on the block.

By the spring of 1998, the first Adopt-a-Block effort was nearing completion. Five of the initial objectives had been accomplished:

1. A functioning block club had been formed.
2. Home improvements had been completed as planned.
3. Beautification of the block had been accomplished, and the block club had adopted a regular schedule of block maintenance.
4. Two homes had been rehabbed and sold.
5. There had been a significant reduction in the levels of crime and illegal activities.

The Adopt-a-Block resulted in other resources being made available. Most significant, the block club and Roseland Homes, working together, requested and were successful in getting the city of Chicago to designate the block as an "Adopt-a-Street." This resulted in access to city resources to improve and maintain the quality of the streetscape.

Although by the summer of 1998 the attention of Roseland Homes was turning to new Adopt-a-Block locations, there were still plans to accomplish the targeted new home construction and a commitment to continue providing social services to residents.

## THE MICHIGAN AVENUE–ROSELAND INITIATIVE

As work was progressing on the Adopt-a-Block, another Roseland community development organization, Neighborhood Housing Ser-

vices of Roseland (NHS), was beginning to pay attention to the deteriorated strip of businesses along Michigan Avenue. The effort to
revitalize Michigan Avenue, organized and "spearheaded" by the
NHS, provided Roseland Homes with a way to enhance its Adopt-a-
Block efforts and to become a part of a coalition of groups working to
bring about overall redevelopment of the community.

The NHS of Chicago had estabtwoished an office in Roseland in 1986
at 110th and Michigan Avenue, just two blocks south of the Roseland
Christian Ministries Center building. Roseland was a logical choice for
an NHS operation because the NHS strategy traditionally has been to
target marginally viable communities that have large numbers of
single-family owner-occupied homes. By assisting Roseland's homeowners in fixing up their homes through the provision of low-interest
loans and technical assistance, the NHS hoped to stem the decline and
return Roseland to being a viable and vibrant community.

The concept for the NHS originated in Pittsburgh in 1968 as a collaborative effort by local residents, foundations, government officials,
and 13 banks and savings and loan associations (Metzger, 1998). Below
market rate loans were made available to homeowners in "high-risk"
areas by creating a loan pool from foundation, government, and bank
funds. The NHS model spread quickly to other communities, and in
1978, the U.S. Congress created the Neighborhood Reinvestment Corporation to sponsor and support NHS organizations and related efforts
in cities throughout the country. Today, there are NHS organizations in
approximately 170 large- and medium-sized cities (Metzger, 1998). The
NHS of Chicago opened its first neighborhood offices in 1975 and by
1979 was operating in five different neighborhoods.[21]

Initially, the Roseland NHS followed the model laid down by other
NHS operations, and they marketed low-interest home improvement loans and technical assistance to Roseland homeowners. However, according to Patricia DeBonnett (personal interview, 1998), the
original director of the Roseland NHS and who after an absence is once
again director, after about a year and a half of operation, it became
apparent that the large number of the HUD-foreclosed, vacant, and
abandoned homes in Roseland was a major problem that, if left unattended, would defeat the traditional NHS efforts to revitalize the
community. So Roseland NHS started helping people buy and rehabilitate the HUD homes. Over the years, DeBonnett estimates that the
Roseland NHS has assisted some 1,000 individuals and has been

involved in the purchase and rehabilitation of more than 200 HUD homes.

Roseland Homes and Roseland NHS shared a common concern, the condition of the Michigan Avenue shopping strip. Both organizations are located on Michigan Avenue. Both have housing developments on or close to Michigan Avenue. Both recognized that unless something was done about Michigan Avenue, their own efforts to upgrade the community through rehabilitating the community's housing could be thwarted.

The NHS took the lead in addressing the condition of Michigan Avenue, perhaps because it had a greater investment in housing near or along the street. In the mid-1990s, the NHS had begun to construct new single-family homes on locations only a block from Michigan Avenue through the use of a new city of Chicago program, "New Homes for Chicago." NHS staff were concerned that it might be difficult to find buyers for these new homes as long as Michigan Avenue remained in such a sorry state. The New Homes for Chicago program, by the way, is the same program that Roseland Homes anticipates using to carry out its new housing construction.

The NHS plans call for the eventual construction of 30 new homes within one block of Michigan Avenue.[22] NHS also has plans to construct a 40-unit apartment building and a senior building, both to be built directly on Michigan Avenue, on land it already owns. Clearly, the condition of Michigan Avenue is important to the NHS.

However, the condition of Michigan Avenue is also important to Roseland Homes because the original Adopt-a-Block is only one block away. Thus, when the NHS began to solicit support and participation in an effort to address the problems of Michigan Avenue, it was not surprising that Roseland Homes was eager to take part.

There had been previous attempts in the early 1990s to address the state of the Michigan Avenue business community, but not much had been accomplished. These earlier efforts encountered what had been a common problem in Roseland, a proliferation of organizations that were unable to reach a consensus as to what needed to be done. Attempts at collaboration had usually ended in failure in Roseland.

But in 1997, the board and staff of the Roseland NHS decided to once again look at Michigan Avenue and attempt to build support for a collaborative revitalization effort. Through funding from the Chicago

Community Trust, the NHS established what would later become known as the Michigan Avenue–Roseland Initiative. The NHS hoped that the initiative would not be seen as belonging to any one group but would be viewed instead as truly a community-wide project. From the beginning, NHS staff were careful to describe their role as "spearheading" the initiative. Operationally, this meant that a committee, consisting of a coalition of community groups, would make the decisions, and the NHS would provide staff for planning and implementing the initiative.

Roseland Homes was a participant in the first committee meeting that brought together an ad hoc group of community stakeholders, including representatives from local banks, residents who had demonstrated past involvement in the community, and local community organizations. Over time, more than 20 organizations and individuals participated as part of the committee.

Maurice Williams says that it was important for Roseland Homes to get involved in the initiative because, in his opinion, Michigan Avenue is critical to all other Roseland revitalization efforts. Calling Michigan Avenue "the face" of the community, he argues that it will not be possible to keep existing quality residents or to attract new ones to the community unless Michigan Avenue is improved.

The ad hoc committee decided that its first task would be to get the Chicago Department of Planning interested in redeveloping Michigan Avenue. If they could attract the city's attention, it might be possible to get it to designate Michigan Avenue as a "commercial redevelopment area." Once so designated, city resources would be available to encourage and support revitalization.

It took 9 months for the committee to convince the city to designate the street a commercial redevelopment area. During this time, the committee developed its own revitalization plan and presented it to the planning consultants hired by the city to prepare the formal designation. The committee members felt that they could not count on the city or its consultants to devise a plan that would be workable, would be acceptable to local stakeholders, and, over time, would be implemented.

Maurice Williams and Roseland Homes became active participants in the planning project, and when the ad hoc committee decided to adopt a more formal structure and become the Michigan Avenue–

Roseland Initiative, Williams became the group's chair. Thereafter, he would work closely on completing and implementing the plan with an NHS staff planner, Angie Marks.

Marks had been hired by the NHS specifically to carry out the Michigan Avenue planning process. She had recently received her master's degree in urban and regional planning at the University of Illinois at Urbana-Champaign, where she had studied under Ken Reardon. Her work with the East St. Louis Project (Reardon, 1997), an advocacy redevelopment effort headed by Reardon, had given her the experience she needed to assist the Michigan Avenue effort.

Chicago State was able to continue its assistance to redevelopment work in Roseland by providing Marks with a student intern. This intern undertook a land use inventory and evaluation of all the properties along Michigan Avenue and, by the end of his internship, had begun to construct a database for eventual use in a Roseland Geographic Information System (GIS).

After achieving the redevelopment designation for Michigan Avenue, the ad hoc committee decided to become a more permanent organization to be able to oversee the implementation of the redevelopment plan. In August 1998, the group was renamed as the Roseland Redevelopment Planning Board. Much work needs to be done before Michigan Avenue becomes a true community asset, but the work that has been done to date has benefited other community revitalization efforts, including those of NHS and Roseland Homes.

## EXPANDING THE ADOPT-A-BLOCK AND THE COLLABORATIONS

By fall of 1997, Roseland Homes was at a crossroads. The first Adopt-a-Block effort was nearing completion, and Michigan Avenue had been designated a commercial redevelopment area. Yet funding for new home construction on the Adopt-a-Block had not been secured, and the rehabilitation of existing properties was proceeding more slowly than had been anticipated. Not wishing to lose the momentum that had carried the organization to this point, Williams decided it was time to move forward and once again expand the work of Roseland Homes. This meant becoming involved in two distinct but related initiatives: entering into a collaborative effort with the Chicago United Way/Crusade of Mercy and expanding the Adopt-a-Block program.

The Chicago United Way/Crusade of Mercy has been a leader among United Ways throughout the country in the area of housing. Since 1990, it has promoted the long-term viability of community development corporations by making "priority" grants for community-based development projects and helping community development corporations (CDCs) to become permanent United Way member agencies (Peterman, 1997). By the mid-1990s, the United Way/Crusade of Mercy had become the third largest source of nonfederal funding for Chicago-area CDCs (Brown, 1997).

Recognizing the need for housing development groups to expand beyond the mere provision of housing, the United Way's Community Development Initiatives Committee (CDIC) had initiated a "collaborative project" that was intended to link CDCs with other United Way agencies in their communities and to provide technical assistance that would give CDCs the skills they needed to better use the services and skills of these agencies. Three CDCs had participated in the first round of the collaborative, and there had been one successful collaboration. One collaboration failed when the CDC itself failed, and the second one never got off the ground due to the CDC's preoccupation with other activities. Despite this somewhat mixed success, the committee decided to solicit proposals for a second group of collaborations.

As a member of the United Way's CDIC, I was aware that the second round of collaborations was being planned, and I mentioned it to Maurice Williams. He was immediately interested, seeing the collaboration as a way to boost Roseland Homes' goal of providing Adopt-a-Block residents with a complete array of social services. After one or two meetings with United Way staff, Roseland Homes submitted a proposal to the CDIC and was accepted as one of three new collaborative partnerships scheduled to begin during the summer of 1998.

Although the Roseland collaborative is just beginning as this chapter is being written, Williams sees it as quite beneficial to Roseland Homes. He gives four reasons why: It gives name recognition of Roseland Christian Homes Corporation's efforts, provides the organization with valuable technical assistance, provides support that can be used to raise funds for the Roseland Christian Homes Corporation, and assists in enlisting support of other organizations in the community.

Roseland Homes is also in the process of expanding the Adopt-a-Block program to three new blocks, 108th and Wabash Avenue (the block immediately to the north of the first Adopt-a-Block), 102nd Place between State Street and Michigan Avenue, and 110th and Edbrook.[23] The kickoff for the first of these three efforts was held in August 1998, and kickoffs for the other two were anticipated by the end of 1998. Williams was hoping that the experience gained with the first Adopt-a-Block would make it possible for Roseland Homes to handle three new blocks at one time. Unfortunately, as of the summer of 1999, neither of the two anticipated Adopt-a-Blocks has been initiated. The primary cause for the delay has been lack of external funding needed to move the project forward.

Chicago State continues to assist Roseland Homes. We plan to supply Roseland Homes with yet another intern who will assist in the United Way collaborative effort. We are also indirectly supporting Roseland Homes by seeking funding to implement one piece of the Michigan Avenue plan. If funded, we will assist in teaching young men from the community to make landscape improvements for existing businesses and in vacant lots along Michigan Avenue. Our partners on this proposed project are Chicago Wilderness, a coalition of Chicago environmental organizations, and yet another Roseland organization, the Soweto Center.

\*   \*   \*

Roseland Christian Homes Corporation is the type of community development organization that Stoecker (1997) has criticized as being too small and having too little activity to be effective in revitalizing the community in which it is located. This criticism would perhaps be valid if Roseland Homes had simply continued its initial program of purchasing and rehabilitating one home at a time in scattered locations. Its expansion, first to one Adopt-a-Block, and later to three more Adopt-a-Blocks, all four of which are within the same general area of Roseland, suggests that the organization has a strategy to affect the community despite its small size.

Furthermore, Roseland Homes has reached out and joined together with other Roseland organizations to undertake tasks that are critical to the redevelopment of the community. The Michigan Avenue–Roseland

Initiative is too big a task for any one Roseland community group to address alone, and collaboration is essential. Roseland Homes' recent move to become a participant in the United Way collaborative project is yet another way the organization is trying to expand its activities by reaching out and using resources that already exist in the community.

Yet despite nearly 20 years of work, first as a task force and later as a CDC, Roseland Homes has only had a modest impact on the direction of the Roseland community. It appears that if it is to have a greater impact, the organization will have to grow and continue to expand its work. New staff will be needed as well as new sources of funding. Difficulties in raising the funds needed to fully implement the first Adopt-a-Block effort hint at problems that could eventually undermine the organization's goals.

Stoecker (1997) may yet be proven right. Currently, Roseland Homes is too much the effort of a single individual, Maurice Williams. Williams will need to bring others into leadership positions, or there will be limits as to how far the organization can expand. What if Williams leaves the organization? Will there be anyone who can continue the work of Roseland Homes, or will much of what has been accomplished evaporate?

This case study suggests that small community development corporations have the potential to act as effective change agents in communities. But it also suggests that they cannot do it alone. Maurice Williams says that he and the leaders of other Roseland organizations understand the need to work together. He says that the forces that brought about the decline of Roseland are still present and strong. But he remains optimistic about the community and believes that through hard work and by collaborating when and where appropriate, Roseland will eventually be revitalized.

## NOTES

1. Chicago State University, where I now teach, is actually located within the formal boundaries of Roseland, although most people think of Roseland as that area about a mile or two south and slightly west of the university.

2. Roseland is one of the few community areas in Chicago where population continued to increase into the 1980s. In 1940, the population was just over 44,000, and in 1980, it

was more than 64,000. The population dropped in 1990 to about 56,500, in part due to considerable housing abandonment and some demolition.

3. At this time, housing discrimination was still legal. The movement of blacks into the northern end of Roseland represented the "natural" growth of the black ghetto southward from neighborhoods further north.

4. Even today, community residents talk fondly about traveling to Michigan Avenue for a day of shopping. Gately's, which finally closed in the 1980s, is particularly singled out as being a "special" place.

5. The 108th Street site in effect was the national headquarters of the Christian Reformed church.

6. By the 1970s, only two of the original eight Christian Reformed churches remained.

7. Patricia DeBonnett, who currently is the director of the Roseland Office of Neighborhood Housing Services, was a travel agent in the Roseland community and handled the travel accounts of the Back to God Hour staff. When hearing that the Back to God Hour planned to move and leave the building empty, she encouraged the organization to find some use for it, fearing the impact that such a large empty building would have on the surrounding neighborhood. She became a member of the original task force that established Roseland Christian Ministries Center.

8. According to DeBonnett, the building had been given to the Christian Reformed church with the legal provision that it was never to be sold. Thus, the church really had only two alternatives: find a new use for the building or let it stand empty.

9. The intern had become disturbed about the housing condition in the community after counting 33 boarded-up and abandoned houses in a 2-block radius of the Roseland Christian Ministries Center.

10. This house was not actually in Roseland. But the task force used the money from the sale of this house to purchase a second house that was in Roseland.

11. "Reverend Tony" van Zanten likes to tell a story about Maurice Williams. He says that one day, back in 1977, when the center was trying to create a "one-basket gym" out of an old parking garage section of the building, he answered a knock at the "gym" door. On opening the door, in rushed several "little" kids. One of them was Maurice. Van Zanten says that from that time on, Maurice Williams was a part of Roseland Christian Ministries Center.

12. Unlike many of Chicago's low-income neighborhoods, Roseland consists primarily of single-family homes, and the homeownership rate is high at 66% of all units. This is significantly higher than the 42% ownership rate for the city as a whole.

13. It is the goal of Roseland Homes to allow families to purchase a home for a monthly payment that is equal to or below the cost of a Chicago Housing Authority rental and to see that no family pays more than 30% of its income for the mortgage payment, taxes, and insurance.

14. The "Adopt-a-Home" program is only one of Roseland Homes' programs. In addition, it trains young men from the community for its "Revive" construction crew, owns and manages 27 apartment units in Roseland, and holds monthly "home buyers club" classes in home maintenance and budgeting. Purchasers of the Adopt-a-Homes must attend the home buyers club meetings to help ensure they will be able to maintain the home in the condition in which they received it.

15. Although this represented a progressive move by Evergreen Bank and its president, it must be pointed out that the partnership arrangement also allowed the bank to

respond positively to its obligations under the requirements of the federal Community Reinvestment Act.

16. I already knew about the Roseland community prior to coming to Chicago State University. While at the University of Illinois at Chicago, a graduate student and I had prepared a planning document called the "Roseland Needs Assessment," which we had developed for several Roseland community leaders. I felt at the time that most, if not all, Roseland community groups were too small to be effective in addressing the community's many problems. Our work on the "assessment" had been supported by a local Roseland-area bank.

17. Some church members of the Roseland Christian Ministries Center also lived on the block.

18. The cost of the public infrastructure improvement was estimated to be $150,000 and would be sought from the city of Chicago. The remainder of the needed funds was to be acquired through foundation grants, donations, and any other source that could be identified.

19. Not all who attended were block residents. At the first formal block club meeting of the 109th and Wabash group, 10 families were represented.

20. Roseland Homes would later use the knowledge gained in forming the block club to develop an information packet on the value of block organizations and how to form one. This packet was sent to households on all 30 blocks where the Roseland Christian Homes Corporation had developments.

21. The Chicago Neighborhood Housing Services (NHS) is one of only a few NHS organizations operating in multiple neighborhoods. The goal of each local neighborhood NHS office in Chicago is to have a stabilizing impact on the community in which it is located, stimulate revitalization, and then leave. Staff of the Roseland NHS recognize that much work yet remains before they can begin thinking about phasing out their office.

22. By fall of 1998, the NHS had completed construction of seven units that it was attempting to sell.

23. All three of these new blocks are within one block of Michigan Avenue. Thus, the Michigan Avenue–Roseland Initiative will have continuing importance to Roseland Homes at least into the immediate future.

# CRITERIA FOR DOING SUCCESSFUL NEIGHBORHOOD DEVELOPMENT

Each of the previous four chapters presented a situation in which a neighborhood and one or more community-based organizations in that neighborhood attempted to bring about a hoped for neighborhood future. None of the efforts undertaken was totally successful, although, with the exception of the first situation (the West DePaul Concerned Allied Neighbors' attempt to address gentrification), some level of success was achieved.

Planners and planning were involved in each situation, although the nature and scope of planning activities undertaken varied from case to case. Given the variety of situations, their ultimate outcomes, and the

153

planning activities engaged in, it may appear to be difficult at first glance to draw from these four cases any general conclusions about either neighborhood development or the role that neighborhood planning can play in assisting neighborhood development efforts. However, on reflection, it can be shown that there are characteristics common to each situation. These common characteristics form the basis for the discussion in this chapter.

## IDENTIFYING COMMON CHARACTERISTICS AND CRITERIA

After the residents of Leclaire Courts successfully achieved their goal of becoming resident managed, as discussed in Chapter 7, I sought and received a small grant from the Woods Charitable Fund to reflect on the resident management process and to evaluate its usefulness as a tool for revitalizing public housing. One of the activities I undertook as part of the grant was a consultation about the resident management process. I asked three experts in the area of housing and community development—Rachel Bratt, Charles Connerly, and Daniel Monti—to join me in Chicago for 2 days. During that time, we talked among ourselves and met with resident leaders and others associated with public housing.

The goal of the consultation, as I outlined it, was to see if we could identify critical elements of a resident management program that, if present, would lead to both physical improvement and resident empowerment. The lively conversations held during the consultation did much to shape my thinking about resident management and greatly influenced the way I began to talk about the subject. These thoughts are perhaps best expressed in the chapter about resident management that I wrote (Peterman, 1993) for the book *Ownership, Control, and the Future of Housing Policy,* edited by R. Allen Hays.

At the time of the consultation, Dan Monti had just completed an evaluative study of 11 resident management corporations throughout the country (Monti, 1989). We used his findings as the starting point for our discussions about what was needed for a successful conversion of conventionally managed public housing to resident management. Monti had found that four conditions were necessary for a successful resident management effort. These conditions, with some slight modifications, are presented in the Hays's book (Peterman, 1993, p. 166).

Over time, I have come to believe that these four conditions have a broader application than applying only to resident management. With proper modification, I believe they are also the conditions necessary for successful neighborhood development of any kind and are applicable to any neighborhood and neighborhood organization. These four criteria, as restated to be more generally applicable, lie at the heart of the arguments about doing neighborhood development in this book and are the key to interpreting the outcomes of the case studies presented in the previous chapters. I have come to believe that each of these criteria must be met for neighborhood development to succeed. If this is true, then an important responsibility of a planner involved in neighborhood development is to pay attention to each of these criteria, seeing, if at all possible, that they are met.

Thus, with thanks, especially to Dan Monti, but also to Rachel Bratt and Charlie Connerly, I propose the following four criteria for successful neighborhood development:

1. Adequate and ongoing monetary resources as well as human technical resources must be available and accessible not only to carry out individual development projects but also to sustain a comprehensive program of neighborhood development or redevelopment.
2. Community development must be demand driven, arising from grassroots community organizing. It cannot be legislated into existence by public officials, no matter how well intentioned.
3. Community leaders must build and maintain strong and direct ties with public officials; technical, legal, and financial experts; and other community organizations and umbrella coalitions of organizations.
4. The relationships between the community and those governmental agencies that have interests in and responsibilities with respect to the community must be neither too friendly nor confrontational. An atmosphere of "creative tension" appears most appropriate.

The first of these criteria speaks to the issue of disinvestment and the role that it plays in creating neighborhood decline. The need to do something about development or redevelopment nearly always comes about because a neighborhood has been disinvested, whether by the private sector, the public sector, or by both. Although disinvestment often means the withdrawal of capital from a neighborhood through redlining by lenders or insurers or through the withdrawal of public services, it can also mean the withdrawal of human resources, such as the human capital associated with both public and private institutions. Efforts to redevelop a community must therefore be mindful of the lack of existing resources. Without bringing resources, both capital and human, to the planning process and to the development strategies, it is unlikely that revitalization efforts will succeed.

The second of these criteria speaks to the issues of community empowerment and control. Revitalization strategies rarely succeed without community involvement. Neighborhood revitalization not accompanied by community empowerment brought about through community organizing and organization almost always results in gentrification and displacement. There is no set structure for community organizations, and the type of organization will vary from neighborhood to neighborhood. Whatever its type, however, community organizations are essential to ensuring that revitalization benefits the existing community and its residents.

The third criterion, as does the first, runs somewhat counter to arguments put forward by proponents of asset-based community development. According to the asset-based model, the resources needed for redevelopment already exist in a neighborhood but are underused. Although I do not deny the existence of neighborhood assets, I argue that assets are not distributed equally throughout all neighborhoods. Poor neighborhoods are not run-down because their residents have failed to apply themselves to making the neighborhood better. Poor neighborhoods are also resource poor. Perhaps a few poor neighborhoods can organize their resources in ways that will allow them to mount effective redevelopment efforts, but most will have to reach beyond the neighborhood boundaries and create linkages with political, technical, legal, and financial experts who possess and can bring resources to bear on a community's needs.

The last criterion is very important and, in some ways, the most controversial of the four. At its heart is the notion of advocacy, community organizing, and the identification of powers external to the neighborhood as "the enemy." It also recognizes, as articulated by Jane Jacobs (1961), notions of the street neighborhoods, as well as Bowden and Kreinberg's (1981) notions of where power lies in a city and just how far neighborhoods are from the real sources of political power.

According to Jacobs (1961), there must be a broker organization between the street neighborhoods and the city government if neighborhoods are to gain access to power. The creative tension referred to in this criterion comes about through the successful act of brokering. My argument is that brokering only succeeds when the process involves creative tension. This implies that the process of revitalization is an ongoing struggle to find a middle ground between the desires and interests of the overall power structure, as well as the desires and interests of those people living in local places.

## ASSESSING THE CRITERIA FOR SUCCESSFUL NEIGHBORHOOD DEVELOPMENT

If the four criteria listed above are as important as I contend, then we should be able to use them to analyze each of the four case studies in Chapters 5 through 8 and to understand why some efforts succeeded but others failed. We should be able to show that when and where neighborhood development was successful, the criteria were met, and when and where there were failures, they were not. If this is true, then the criteria can be thought of as a set of guidelines to follow when attempting to plan and bring about neighborhood development. These guidelines will be useful to us whether we are advocate planners, public planners focusing on neighborhoods, foundation representatives interested in encouraging neighborhood development, community organizers, or neighborhood residents.

In the West DePaul situation, presented in Chapter 5, none of the four conditions was met, and the effort to forestall or cope with gentrification failed. The community group, Concerned Allied Neighbors (CAN), was small, was poorly organized, lacked resources, and had few if any community linkages, either internally or externally. The changes that occurred in the community's makeup as gentrification began to take hold in the neighborhood further destabilized the organization and led to its metamorphosis to an organization that favored rather than opposed gentrification. Finally, because the group's leaders were reluctant to directly confront the forces responsible for neighborhood change, there was no opportunity for creative tension to develop between the organization and either city officials who supported gentrification or the "gentrifiers" who were causing it.

In the section situation, when South Armour Square was faced with the imminent construction of a baseball stadium, there initially was no community organization to represent the local residents. But at the request of some residents, a community organizer was brought in and was able to quickly create a coalition of community residents for the purpose of fighting the community's destruction. The organizer was also able to bring other resources into the community, such as the Voorhees Center, and was more or less successful in forging links with groups from outside the community who became allies in the fight to save the neighborhood.

But at no time was there a creative tension between the neighborhood and those with power over its existence. The city, state, and the White Sox were determined to have their way and were totally uninterested in dealing with the resident organization. Instead of dealing with the coalition of groups, the city of Chicago chose to secretly negotiate a settlement with a small group of homeowners. This act undercut the efforts of the community organization, and after the settlement was reached and announced, the coalition rapidly disintegrated.

Nonetheless, the remaining elements of the community coalition, especially the residents at the Wentworth Gardens public housing development, were able to regroup and to carry on the fight. This eventually resulted in positive improvements to the remaining neighborhood.

It is not surprising that the Wentworth residents were able to continue because their community was already organized. Although the proposal to build the stadium was undoubtedly the most serious threat to Wentworth that the residents had ever faced, they were already used to working together and were able to put together a coherent and sustainable effort to oppose it. Furthermore, Wentworth residents already had developed links to resources and with individuals and groups from outside the neighborhood, and they were able to continue using these resources and links after the battle over the stadium was lost. Thus, the failure to stop the stadium did not stop the Wentworth residents who continued their efforts for community improvement.

In the third situation, the attempt at Leclaire Courts to create the first resident-managed public housing in Chicago was initially successful because planners, social workers, lawyers, housing experts, and others all worked together with a core of residents to create conditions that would allow community development to occur. The establishment of a resident organization was seen as an early and important task. Committees of outside experts were formed, and this led to legal, financial, and developmental assistance for the developing community. The local alderman and congressional delegates were kept informed of the project, and their support for the effort was solicited. Funds were raised to do some of the planning and for training the staff and board of the corporation. And a creative tension between the resident organization and the Chicago Housing Authority (CHA) was present from almost the beginning.

But after the resident corporation took control of Leclaire, the creative tension slowly but surely ebbed, especially after Vince Lane became the chairman of the CHA. He embraced the concept of resident management and encouraged the resident management corporation to work more closely with the CHA. As the resident leaders began to rely more on the friendly officials at the CHA, they began to turn away from many of their advisers. One by one, their links to the external community were severed, and they came to be seen as more of an arm of the CHA than an independent corporation. Over time, the resident management effort stalled and then ultimately failed.

In the fourth and final situation, Roseland Christian Homes Corporation, a small community development corporation (CDC) in Roseland, was presented as attempting to revitalize the neighborhood through housing rehabilitation and homeownership. To be successful, Roseland Homes has had to expand its influence and has done so by reaching beyond itself to obtain resources that it does not possess, including funds from external "partners" that are used to finance the renovation and purchase of individual homes.

The creation and ongoing support of an organized and active block club were critical to the initial success of the Roseland Homes' Adopt-a-Block effort. Ongoing activities of the block club are also seen as necessary to ensure that the physical improvements made during the Adopt-a-Block effort will last and result in continuing block development.

Roseland Homes also has recognized the importance of using resources that exist within the community, as is evidenced by its association with Neighborhood Housing Services (NHS) and the Michigan Avenue–Roseland Initiative. It also has recognized the importance of external links and has developed several, including its ongoing association with the Evergreen Bank and with sponsoring "partner" churches.

It is less clear how the notion of creative tension applies to the Roseland Christian Homes Corporation. The organization has faced very little opposition to its efforts to renovate and sell individual homes and to its Adopt-a-Block effort, unless you count the displeasure of local drug dealers and other "characters" for whom the redevelopment efforts are threatening. But as Roseland Homes expands its efforts through collaborations with other organizations, it may ultimately be faced with opposition.

If we expand the scope of this fourth case study beyond a focus on the Roseland Christian Homes Corporation, the situation with respect to creative tension takes on a different cast. The Michigan Avenue–Roseland Initiative has clearly involved the application of creative tension. Support from the city of Chicago was required to designate Michigan Avenue as a "commercial redevelopment area." The ad hoc committee, spearheaded by the NHS, sought city support but did not wait for city planners to propose a redevelopment plan. Instead, the committee developed its own plan, which was then presented to the city's representatives. Thus, the ad hoc committee neither capitulated to the city nor totally opposed it. In addition, some of the initial efforts to implement the redevelopment plan by the ad hoc group, now reconstituted as the Roseland Redevelopment Planning Board, are leading to conflicts between the group and some local merchants. Whether this conflict can be a source of creative tension or whether it leads to further conflict and opposition to the redevelopment efforts remains yet to be seen.

We could analyze each of the case studies in further detail, but it seems from what we have done so far that the four criteria for neighborhood development are useful in explaining the outcomes of each situation. Thus, they can be considered the least important components of any neighborhood development effort. I like to think of them as the necessary if not always sufficient conditions.

# 10

# DOING PLANNING FOR GRASSROOTS NEIGHBORHOOD DEVELOPMENT

$B$y now it should be obvious to any observant reader that when neighborhood planning is done for the purpose of stimulating urban development, it differs from what is normally thought of as traditional planning. This is the case whether the planning is done from the bottom up or if it is top-down subarea planning. But even after exploring how planning assisted each of the efforts described in Chapters 5 through 8, it still may not be clear as to just how it differs or, more important, just what development-oriented neighborhood planners do.

Based on many years of working on development projects with community organizations, community organizers, and neighborhood advocates, I have arrived at conclusions as to some general characteristics of neighborhood planning. They are discussed in the following paragraphs.

Neighborhood planners accept Davidoff's (1965) assertion that planning is not a value-free activity. This does not mean that neighborhood planning is subjective. Rather, it means that although neighborhood planners try to be objective in their work, they also try to understand how their values affect the choices they make in both choosing how to study an issue and how to interpret the results of their work. They are also explicit about the values underlying their work in both their written and oral pronouncements.

In general, neighborhood planners do not see their scope of work as being limited to the technical procedures traditionally associated with the rational planning model. Nor do they see their role as simply makers of plans. This does not mean, however, that they avoid making plans or using "rational" planning techniques.

For the most part, as we have seen in the case studies, neighborhood planning does not consist solely of developing neighborhood plans. Neighborhood planning is about addressing the inequities that result from poverty and neighborhood disinvestment. Such inequities are best addressed by working to change policies and programs that have caused uneven development or by attempting to create reinvestment opportunities in blighted neighborhoods. Often, the most important thing a neighborhood planner can do is to use his or her knowledge and skills to help neighborhood residents and others understand the impacts of urban policies and programs and to identify and help obtain those resources a community needs to begin the processes of revitalization, redevelopment, and reinvestment.

At times, actual plans will be needed to articulate a community's vision for its future, and if a plan is needed, then it is appropriate for a neighborhood planner to facilitate the planning process. Any plan produced, however, should be thought of as merely another tool for enabling a community to realize its goals and objectives and should not be thought of as being the neighborhood planner's primary task.

Neighborhood concerns, problems, issues, and goals are rarely neatly arranged so they can be handled by undertaking a classic planning procedure or by producing a traditional planning study. Neighborhood planners need a broad portfolio of skills, some of which are not commonly thought to be part of a planner's toolkit.

In his guide to neighborhood planning subarea style, Bernie Jones (1990) says that neighborhood planning must be "democratic planning," and even though he cautions neighborhood planners to not get too involved in the neighborhoods in which they work, he says that planners need to be capable of extending beyond their traditional roles. In addition to the usual technical skills—such as map making, performing quantitative analysis, and having knowledge of planning and urban theories—Jones argues that planners who do neighborhood planning need to be conveners, facilitators, scribes, and even gophers, seeing that everything needed for a community meeting has been done ahead of time. Planners whose education did not include "people skills," Jones states, have been cheated.

There are three key characteristics associated with any community-based planning process. First and foremost, as depicted in the previous chapters, neighborhood planning is a collaborative process, involving planners, community organizers, other experts, and the neighborhood. Planners bring certain skills and knowledge to the process, but they must recognize that they do not have all of what is needed to move any neighborhood development project forward. Other "experts," such as lawyers and lenders, may be needed to move the planning process forward. Community organizers are critical to bringing together the neighborhood residents and helping to devise strategies to move from plans to action.

But most important planners must recognize that their work is meaningful only when it is done collaboratively with the neighborhood. This means recognizing that planners do not have all the answers, nor do they have all the knowledge as to the nature of the "problem." Neighborhood residents bring their own knowledge and skills to the planning and development process, and these complement the knowledge and skills of the so-called "experts."

A true collaborative planning process is one in which all parties participate as equals. Planners must respect and value the opinions and ideas of others, even when they are inconsistent with their own points of view. If neighborhood planners are also to be neighborhood advocates, they must listen to neighborhood residents, understand what is important to them, and try as much as possible to address their concerns.

Because the neighborhood planning process is a collaborative one, the planner also is free to express his or her concerns and points of view.

A planner who feels that what the residents believe they want with regard to any project is not in their best interests has an obligation to express his or her concerns, try to persuade the community that their ideas are inappropriate, and try to convince them that his or her ideas are better. In the end, however, it is the neighborhood that has the final say.[1] Whenever it becomes clear that the values of the planner and the community are at odds, the planner has both an obligation and responsibility to make these differences known and, if necessary, to withdraw from the project.

The second key characteristic is that neighborhood planning must be open and transparent. What a planner does, how he or she chooses a methodology, and how conclusions are reached should not be a secret. Neighborhood residents have the right to see and understand all that the planner is doing. This may require the planner to think through what techniques and procedures are best suited and how they are to be explained. "Black box" approaches should not be used. Properly instituted, neighborhood planning is an educational process, and the planner is a teacher. The ultimate goal is to teach the community and its leaders how planning and development are done so that when the planner leaves the community, the work will continue. This, for the planner, is the ultimate act of empowerment.

Finally, all neighborhood planning should be community driven. This is what it means for planning and development to be community based. Residents acting collectively can express what they feel to be the important issues facing the community, and they can identify their concerns, hopes and fears, and aspirations for the future. The planner's responsibility is to listen carefully, try to understand as much as he or she can, and help the community achieve its goals when and where possible. Whenever neighborhood planning loses sight of the neighborhood's agenda and begins to address the planner's agenda, meaningful neighborhood planning has ended.

Marie Kennedy, of the University of Massachusetts at Boston, argues that true community-based planning is a transformative and empowering process combining material development with the development of people. It should, she states, "leave a community not just with more immediate 'products' . . . but also with an increased capacity to meet future needs" (Kennedy, 1996, p. 12). According to Kennedy, the measures of the success of such a transformative and empowering community planning process should be the following:

- the control of development being increasingly vested in community members;
- increasing numbers of people moving from being an object of planning to being a subject;
- increasing numbers of confident, competent, cooperative, and purposeful community members;
- people involved in the planning process gaining the ability to replicate their achievements in other situations; and
- movement toward the realization of the values of equity and inclusion.

In light of the above discussion, it still remains difficult to provide a precise outline of the neighborhood planning process or to make a list of what tools and procedures neighborhood planners use. This is probably disconcerting to planning traditionalists who feel that it is necessary to have identifiable theories and procedures. Although wishing to avoid becoming entangled in any of the debates about the current state of planning theory and its relationship or lack of relationship to planning practice (see, e.g., Hall, 1989), let me simply suggest that neighborhood planning carried out as outlined in this book is more consistent with the ideas of those who support postmodern notions of planning than it is with the ideas of those who believe that planning should follow the tenets of traditional rationalism (see Milroy, 1991).

## DOES IT MAKE SENSE TO FOCUS ON NEIGHBORHOOD PLANNING AND NEIGHBORHOOD DEVELOPMENT?

There is a subtext running throughout this book that, although critical to the issues and actions presented in the case studies of Chapters 5 through 8, is not directly addressed in any of the four chapters. The subtext centers on two basic questions: Are neighborhoods real entities with meaning in the modern city, and does it make sense to focus a significant amount of planning and development energy at the scale of geography that we commonly refer to as the neighborhood level? Several years ago, the answers were obvious to me. Neighborhoods exist and are important elements of urban life. Thus, planning should pay attention to them. These answers, almost statements of faith, are no longer as satisfying to me as they once were.

In Chapter 4, I referred frequently to the book *Street Signs Chicago*, written by Charles Bowden and longtime community activist and organizer Lew Kreinberg (1981). The subtitle of this book is *Neighborhood and Other Illusions of Big-City Life*, and in it the authors assert that neighborhoods are confused with European villages and that they are merely "designations and handles for the schemes of politicians, planners and political scientists . . . but they are not a commitment of families to a place through time; nor have they ever been" (Bowden & Kreinberg, 1981, p. 75). The emphasis on neighborhoods, these authors claim, is a ruse perpetrated by those in power to obscure the very issue of power—who has it and who does not.

The question about the reality neighborhoods was first addressed in Chapter 2, in which I suggested that *retrospective modernism,* viewing the past through criteria applicable only in the present, was likely to be responsible for the belief that neighborhoods have been central elements of cities down through history. In that chapter, I also suggest that the modern notion of the neighborhood probably originated with the thinking of social reformers and idealists such as Ebenezer Howard and George Pullman, and its ideal form more closely resembles the upscale subdivisions of late 19th and early 20th century British and U.S. cities than it does the "village in the city."

I also pointed out that urban sociologists, such as Gerald Suttles (1972) and Herbert Gans (1991), do not see neighborhoods as ubiquitous elements of the urban landscape. Although both recognize that in some parts of the city, places we call neighborhoods do exist, they are not the organizing principle around which cities are organized. Indeed, they tend to be the exception—the places that are different from the rest of the city.

In Chapter 3, after discussing issues of empowerment and community organization, I raise the question of whether attempting to do community development at the local level is an appropriate or effective pursuit. I do this by presenting a discussion of Randy Stoecker's (1997) stinging critique of the community development movement. Stoecker argues that community development corporations (CDCs), as they are currently structured, are too small and too narrowly focused geographically to be capable of actually revitalizing communities. Furhermore, he contends that focusing on neighborhood-based re-development plays into the hands of those city officials and their de-veloper friends who are the real power in cities.

Focusing on neighborhoods and neighborhood planning may not only be ineffective, but it may also be detrimental to the quality of urban life. In a powerful essay written more than two decades ago, Richard Sennett (1976) argued that the urban planning profession was guilty of atomizing the city through its emphasis on physical organiza-tion, land use, and zoning and that this has resulted in creating what essentially are false communities. Through the development and pro-motion of a land use system in which each class and each race has its own distinct place to live from the central city to the suburbs, planners have isolated like groups of people and have fostered a compulsive search for community life. Rather than achieving community, this search has led to further isolation. Isolating like people in space, in what are in reality ghettos, leads to fear of the unknown, fear of the out-side, and attempts to exclude others from the community. As a result, urban communities have become inward looking, fearful, and defen-sive. Public life of the community at large has been sacrificed for the safety of the community of the neighborhood.

If neighborhoods do not really exist except in limited cases and if focusing on community development is both misplaced and results in isolation, fear, and loss of overall community, why write a book that gives advice to people about doing neighborhood planning and devel-opment?

Neighborhoods, or at least those places we call neighborhoods, are where we, as urban dwellers, live. Although we may work elsewhere, shop elsewhere, use the entire metropolitan area for our network of friends and associates, and travel throughout the region for our leisure time activities, we still spend much of our life close to the places we call home. It is hoped that we still have local places where we can work and shop, parks and other places where we can relax from the hectic life of the urban scene, and people who we call our neighbors. The quality of our home turf depends not just on the quality of our little piece of it we own or rent but also on what surrounds it. Paying attention to neighborhoods means paying attention to the local environment, providing opportunities for human interaction and growth.

It would certainly be a mistake, however, to argue that all or even the bulk of urban planning should focus on geographical places called neighborhoods. Nor would it be particularly productive to focus on neighborhoods at the expense of the broader regions of the city. Neighborhood-based community development, as carried out through community-based initiatives, should be thought of as only one element of any comprehensive strategy intended to foster urban revitalization. Other strategies, focused at other geographical scales, are necessary. Without also attending to economic, social, and political issues that are of concern to a city and to an entire metropolitan region as a whole, efforts to promote development at the local neighborhood level are likely to falter and result in suboptimal outcomes. Even when they are successful, they are unlikely to be sustainable over time. An unhealthy city will have few, if any, healthy neighborhoods.

Let me give an example from my own work of the links between neighborhood issues and more comprehensive urban issues. Along with the neighborhood work that is the topic of this book, I have spent much of my professional career addressing issues of fair housing, housing and lending discrimination, and the creation and maintenance of diverse neighborhoods (see Nyden, Maly, Lukehart, & Peterman, 1998). These are issues of fairness, equal opportunity, and justice, and they cut across neighborhood boundaries. Neighborhoods that contain people who discriminate or unscrupulous real estate agents who are panicking residents to flee are part of the urban problem, not part of the solution. Although issues of fair housing, discrimination, and diversity sometimes focus on the problems of a particular neighborhood, the overall problems associated with these issues cannot be solved solely, if at all, at the neighborhood level.

Federal legislation and aggressive actions on the part of citywide or regional fair housing organizations have been necessary to ensure that all people, wherever they live, have an equal chance to live, study, work, and recreate free from the limitations imposed by discriminatory practices. Issues relating to race and class are intertwined with the issues of neighborhood, but they are issues that play themselves out at the levels of cities, regions, and even the nation. Ignoring issues, such as

fair housing, and arguing that they are not relevant at the neighborhood level result in the segregation of low-income and minority households in disinvested neighborhoods where no amount of
neighborhood planning or local community development can be particularly effective (Massey & Denton, 1993).

To cite another example, it will not be possible to solve the problems
of neighborhood inequalities without also addressing issues of
regional inequalities, and thus the reawakened interest in regionalism
should be viewed as positive by proponents of neighborhood development. The current inequalities in our educational system, which are
often barriers to the recreation and maintenance of sustainable central-
city neighborhoods, are just one example of where the effects are often
local, and the problem is of a regional nature. Other issues of critical
importance to neighborhoods but best addressed regionally include
the deconcentration of public housing and poverty, welfare "reform,"
public transportation, and environmental quality.

## MOVING AHEAD: EXPANDING THE MEANING
## OF NEIGHBORHOOD DEVELOPMENT

Recent studies and writings in the area of neighborhood and community development have begun to suggest that the traditional focus
on housing, economic development, and community safety may be too
narrow and that a broader, more comprehensive, and more interconnected approach may be necessary. The Chicago United Way/Crusade
of Mercy has been undertaking triennial assessments of a variety of
social needs throughout the city, including an assessment of community development needs assessments (Brown, 1997) for many years. A
review of the past few community development assessments demonstrates the changing character of the discussion about community
development.

Each assessment takes almost a year to complete and is done by con-
vening a committee of experts selected from the field that each specific
assessment addresses. The committee is responsible for developing its
research agenda, finding a way to come to consensus about needs, and
preparing the final assessment document. I have been associated with

the needs assessment for community development since 1990, and in 1997-1998, I chaired the assessment committee.

In 1992, the committee wanted to emphasize that no one aspect of community development could stand alone and that important linkages existed between individual components, especially housing and economic development. We wanted to add an introductory chapter in which we tried to show diagrammatically how all the components were connected. This was followed, however, by four separate chapters, each of which related to a discrete component of development: housing, economic development, safety, and the arts.

Those of us on the 1995 community development needs assessment committee who worked directly on its housing component were concerned that past assessments had called for additional low- and moderate-income units in Chicago but had more or less ignored the neighborhoods in which these units would be placed. We were faced with a growing concern that Chicago's community development corporations were hardly making any progress in revitalizing their neighborhoods and that some of them were in trouble from having buildings in neighborhoods where respectable tenants would not live. Thus, buildings with rehabilitated units often were going unrented, and when tenants were found, they often brought their economic and social problems with them, which was placing stresses on the CDCs that were struggling to keep their buildings viable. We concluded in our report that a successful housing program would not only have to address the need for more low-cost units, but it would also have to address the social and economic needs of the tenants as well as of the residents of the community. Housing development, the report concluded, cannot succeed in neighborhoods that are themselves badly in need of improvement.

The 1998 Chicago United Way/Crusade of Mercy community development needs assessment, which I chaired, went yet another step. In our final report (United Way Needs Assessment Committee for Community Development, 1998), we stated that the two most important issues affecting community development in Chicago are the lack of "citizen engagement" and the need for "youth development." Although committee members, most of whom were involved in the nuts-and-bolts activities associated with neighborhood development, recognized the need for continuing efforts to provide housing, jobs, and safe environments throughout Chicago, they felt that this was insufficient. Our conclusion was that community development should really be about *community building*—the creation of viable, healthy, and constructive communities, both place-based communities and communities of interest. Thus, in the report, we tried to make the case that unless the residents of Chicago's many communities become involved in the governance of the community and unless attention is given to preparing the next generation for viable "citizenship," any and all efforts to physically revitalize the city's neighborhoods are unlikely to succeed.

If we reflect on the four case studies presented in this book, we see that the basic neighborhood issue addressed in each situation is really community building. Whether the object was attempting to stave off gentrification as it was in West DePaul or creating resident-managed public housing as it was at Leclaire Court, the hoped for end result was a viable and effective community. Improvements in the physical housing stock or the neighborhood's infrastructure were at best a means for assisting community-building efforts. That is why community organization is so important and is one of the four elements necessary for neighborhood development to succeed.

Community organization, however, is not the only mechanism for community building. Linkages with the broader community and external resources help local neighborhoods ensure that not only are their development efforts successful, but they also create a network of local places that can connect the smaller neighborhoods with more comprehensive efforts to build and sustain community. Linkages help to limit the negative consequences of neighborhood isolation (Sennett, 1976).

I have implied that the neighborhood model put forward by Jane Jacobs (1961) is a more appropriate model for neighborhood development than the classical planning approach of the neighborhood as village in the city. Jacobs understood the importance of linking the small powerless street neighborhoods to the powers of the city through what she called districts. These district organizations—what we would today call community organizations and CDCs—are ways in which local people can come together locally but also be tied to a broader base of resources and expertise. Jacobs's message is restated in the United Way's 1998 community development needs assessment and is demonstrated in the case studies in this book.

So neighborhood planning and development cannot be isolated and inwardly focused. Planners engaged in neighborhood planning and development and community-based developers must continually bear in mind that what they do or attempt to do is important not only for the local neighborhood but also for the larger community. By thinking about neighborhood development on the larger scale and by linking the local with the larger community, neighborhood development can become an effective tool for not only revitalizing local places but also for building from the ground up an effective strategy for urban development, one that will benefit us all.

## NOTE

1. I frequently point out to members of neighborhood organizations with whom I am working that at the end of the day, I will go home to my neighborhood and can, if I wish, ignore what happens or does not happen in their neighborhoods. They, however, cannot. I say this to help them understand that they must take charge of the planning process and be satisfied with its goals and directions. They have to live with anything we accomplish, but I can move on to my next project.

# REFERENCES

Advisory Council on the Chicago Housing Authority. (1988). *New strategies, new standards for new times in public housing.* Chicago: Author.

Alinsky, S. D. (1972). *Rules for radicals: A practical primer for realistic radicals.* New York: Vintage.

Arnstein, S. R. (1969). A ladder of citizen participation. *Journal of the American Institute of Planners, 8,* 216-224.

Austin, M. J., & Betten, N. (1990). *The roots of community organizing, 1917-1939.* Philadelphia, PA: Temple University Press.

Bowden, C., & Kreinberg, L. (1981). *Street signs Chicago: Neighborhood and other illusions of big-city life.* Chicago: Chicago Review Press.

Bowley, D. (1978). *The poorhouse: Subsidized housing in Chicago, 1895-1976.* Carbondale: Southern Illinois University Press.

Bradford, C. (1979). Financing home ownership: The federal role in urban decline. *Urban Affairs Quarterly, 14,* 313-235.

Bratt, R. G. (1989). *Rebuilding a low-income housing policy.* Philadelphia, PA: Temple University Press.

Bratt, R. G. (1991). Mutual housing: Community-based empowerment. *Journal of Housing, 48,* 173-180.

Bratt, R. G. (1997). CDCs: Contributions outweigh contradictions: A reply to Randy Stoecker. *Journal of Urban Affairs, 19,* 23-28.

Bratt, R. G., & Keyes, L. C. (1997). *New perspectives on self-sufficiency: Strategies of nonprofit housing organizations.* Medford, MA: Department of Urban and Environmental Policy, Tufts University.

Bratt, R. G., Vidal, A. C., Schwartz, A., Keyes, L. C., & Stockard, J. (1998). The status of nonprofit-owned affordable housing: Short-term success and long-term challenges. *Journal of the American Planning Association, 64,* 39-51.

175

Brazier, A. M. (1969). *Black self-determination: The story of the Woodlawn Organization.* Grand Rapids, MI: William B. Herdman.

Breitbart, M. (1974). Advocacy planning. In R. E. Kasperson & M. Breitbart (Eds.), *Participation, decentralization, and advocacy planning* (Commission on College Geography Resource Paper No. 25, pp. 41-55). Washington, DC: Association of American Geographers.

Brown, C. (1997, June). *The United Way/Crusade of Mercy's housing initiative: Changing the way United Way views housing.* Paper presented at Housing in the 21st Century: Looking Forward, a meeting of the International Sociological Association Committee on Housing and the Built Environment, Arlington, VA.

Burgess, E. W. (1925a). Can neighborhood work have a scientific basis? In R. E. Park, E. W. Burgess, & R. D. McKenzie (Eds.), *The city* (pp. 142-155). Chicago: University of Chicago Press.

Burgess, E. W. (1925b). The growth of the city: An introduction to a research project. In R. E. Park, E. W. Burgess, & R. D. McKenzie (Eds.), *The city* (pp. 47-62). Chicago: University of Chicago Press.

Chandler, M. O. (1991). What have we learned from public housing resident management? *Journal of Planning Literature, 6,* 136-143.

Checkoway, B. (1984). Two types of planning in neighborhoods. *Journal of Planning, Education, and Research, 3,* 102-109.

Chicago Fact Book Consortium. (1995). *Local community fact book: Chicago metropolitan area 1990.* Chicago: Academy Chicago Publishers.

Clavel, P., & Wiewel, W. (Eds.). (1991). *Harold Washington and the neighborhoods: Progressive city government in Chicago, 1983-1987.* New Brunswick, NJ: Rutgers University Press.

Cleveland City Planning Commission. (1975). *Cleveland policy planning report.* Cleveland, OH: Author.

Davidoff, P. (1965). Advocacy and pluralism in planning. *Journal of the American Institute of Planners, 31,* 331-338.

Dedman, B. (1988, May 1-4). The color of money. *Atlanta Journal/Constitution.*

DeGiovanni, F. (1983). Patterns of change in housing market activity in revitalizing neighborhoods. *Journal of the American Planning Association, 49,* 22-39.

de Tocqueville, A. (1945). *Democracy in America.* New York: Vintage. (Original publication 1835)

Drier, P. (1996). Community empowerment strategies: The limits and potential of community organizing in urban neighborhoods. *Cityscape: A Journal of Policy Development and Research, 2,* 121-159.

Federal Housing Administration (FHA). (1936). *Planning neighborhoods for small houses* (Technical bulletin no. 5). Washington, DC: Author.

Ford, L. R. (1991). A metatheory of urban structure. In J. F. Hart (Ed.), *Our changing cities* (pp. 12-30). Baltimore, MD: Johns Hopkins University Press.

Gans, H. J. (1962). *The urban villagers: Group and class in the life of Italian-Americans.* New York: Free Press.

Gans, H. J. (1991). *People, plans and policies.* New York: Columbia University Press.

Garreau, J. (1991). *Edge city: Life on the new frontier.* New York: Doubleday.

Giloth, R. (1988). Community economic development: Strategies and practices of the 1980s. *Economic Development Quarterly, 2,* 343-350.

Goldsmith, W. W., & Blakely, E. J. (1992). *Separate societies: Poverty and inequality in U.S. cities.* Philadelphia, PA: Temple University Press.

Gunner, J. (1991). *South Armour Square community: A case study of displacement.* Unpublished master's project, University of Illinois at Chicago.

Hall, P. (1989). The turbulent eighth decade: Challenges to American city planning. *Journal of the American Planning Association, 55,* 275-282.

Henig, J. R. (1981). Community organizations in gentrifying neighborhoods. *Journal of Community Action, 1*(2), 45-55.

Hirsch, A. (1983). *Making the second ghetto: Race and housing in Chicago 1940-1960.* Cambridge, UK: Cambridge University Press.

Howard, E. (1945). *Garden cities of tomorrow.* London: Fisher and Faber. (Original publication 1902)

ICF, Inc. (1992). *Evaluation of resident management in public housing.* Washington, DC: Office of Policy Development and Research, U.S. Department of Housing and Urban Development.

Jacobs, A. B. (1980). *Making city planning work.* Washington, DC: American Planning Association.

Jacobs, J. (1961). *The death and life of great American cities.* New York: Vintage.

Janke, D. A. (1996). *Roseland Christian ministries: "I saw the Holy City" 1997-2002.* South Holland, IL: Covenant House Ministries.

Janowitz, M. (1952). *The community press in an urban setting: The social elements of urbanism.* Chicago: University of Chicago Press.

Jones, B. (1990). *Neighborhood planning: A guide for citizens and planners.* Chicago: Planners Press.

Jones, M. (1925). *The autobiography of Mother Jones.* Chicago: Charles H. Kerr & Company.

Joravsky, B. (1987). Comiskey neighbors organize for survival. *The Neighborhood Works, 10*(8), 1, 15-17.

Joravsky, B. (1988, April 22). The stadium game: Who loses if the White Sox win? *The Reader.*

Kain, J. (1968). Housing segregation, Negro employment, and metropolitan decentralization. *Quarterly Journal of Economics, 82,* 175-197.

Keating, W. D. (1989, February/March/April). The emergence of community development corporations: Their impact on housing and neighborhoods. *Shelterforce,* pp. 8-14.

Keating, W. D. (1997). The CDC model of urban development: A reply to Randy Stoecker. *Journal of Urban Affairs, 19,* 29-33.

Kennedy, M. (1996). Transformative community planning: Empowerment through community development. *Planners Network, 118,* 12-13.

Knoke, D., & Woods, J. R. (1981). *Organized for action: Commitment in voluntary associations.* New Brunswick, NJ: Rutgers University Press.

Kotler, M. (1969). *Neighborhood government: The local foundations of political life.* New York: Bobbs-Merrill.

Krumholz, N. (1982). A retrospective view of equity planning: Cleveland 1969-1979. *Journal of the American Planning Association, 48,* 163-174.

Krumholz, N., & Forester, J. (1990). *Making equity planning work: Leadership in the public sector.* Philadelphia, PA: Temple University Press.

Kunstler, J. H. (1993). *The geography of nowhere: The rise and decline of America's man-made landscape.* New York: Simon & Schuster.

Langdon, P. (1994). *A better place to live: Reshaping the American suburb.* Amherst: University of Massachusetts Press.

Leachman, M., Nyden, P., Peterman, W., & Coleman, D. (1998). *Black, white, and shades of brown: Fair housing and economic opportunity in the Chicago region.* Chicago: Leadership Council for Metropolitan Open Communities.

Logan, J. R., & Molotch, H. L. (1987). *Urban fortunes: The political economy of place.* Berkeley: University of California Press.

Manpower Demonstration Research Corporation. (1981). *Tenant management: Findings from a three year experiment in public housing.* Cambridge, MA: Ballinger.

Massey, D. S., & Denton, N. A. (1993). *American apartheid: Segregation and the making of the underclass.* Cambridge, MA: Harvard University Press.

Mayer, N. S., & Blake, J. L. (1981). *Keys to the growth of neighborhood development organizations.* Washington, DC: Urban Institute Press.

McIlwain, C. H. (1936-1937). The historian's part in a changing world. *American Historical Review, 42,* 212.

McKenzie, R. D. (1925). The ecological approach to the study of the human community. In R. E. Park, E. W. Burgess, & R. D. McKenzie (Eds.), *The city* (pp. 63-79). Chicago: University of Chicago Press.

Metzger, J. T. (1996). The theory and practice of equity planning: An annotated bibliography (Council of Planning Librarians bibliography no. 329). *Journal of Planning Literature, 11,* 112-126.

Metzger, J. T. (1998). Neighborhood Reinvestment Corporation. In W. van Vliet (Ed.), *The encyclopedia of housing* (pp. 392-393). Thousand Oaks, CA: Sage.

Mier, R. (Ed.). (1993). *Social justice and local development policy.* Thousand Oaks, CA: Sage.

Milroy, B. (1991). Into postmodern weightlessness. *Journal of Planning Education and Research, 10,* 181-187.

Monti, D. J. (1989). The organizational strengths and weaknesses of resident managed public housing sites in the United States. *Journal of Urban Affairs, 11,* 39-52.

Morris, D., & Hess, K. (1975). *Neighborhood power: The new localism.* Boston: Beacon.

Mumford, L. (1961). *The city in history: Its origins, its transformations, and its prospect.* New York: Harcourt Brace.

Naparstek, A. J., & Cincotta, G. (1976). *Urban disinvestment: New implications for community organization, research, and public policy.* Washington, DC: National Center for Urban Ethnic Affairs.

National Association of Neighborhoods. (1979). *National neighborhood platform.* Washington, DC: Author.

National Congress for Community Economic Development. (1989). *Against all odds: The achievements of community-based development organizations.* Washington, DC: Author.

Nelson, K. P. (1994). Whose shortage of affordable housing? *Housing Policy Debate, 5,* 401-441.

Nyden, P., Maly, M., Lukehart, J., & Peterman, W. (1998). Neighborhood racial and ethnic diversity in U.S. cities and overview of the 14 neighborhoods studied. *Cityscape, 4*(2), 1-28.

Park, R. E., Burgess, E. W., & McKenzie, R. D. (Eds.). (1925). *The city.* Chicago: University of Chicago Press.

Perin, C. (1977). *Everything in its place: Social order and land use in America.* Princeton, NJ: Princeton University Press.

Perry, C. A. (1929). *The neighborhood unit: Regional plan of New York and its environs, VII.* New York: Regional Plan Association.

Peterman, W. (1993). Resident management and other approaches to tenant control of public housing. In R. A. Hays (Ed.), *Ownership, control, and the future of housing policy* (pp. 161-175). Westport, CT: Greenwood.

Peterman, W. (1996). The meanings of resident empowerment: Why just about everybody thinks it's a good idea and what it has to do with resident management. *Housing Policy Debate, 7,* 473-490.

Peterman, W. (1997, June). *Neighborhoods, affordability, and low income housing policy in Chicago, Illinois.* Paper presented at Housing in the 21st Century: Looking Forward, a meeting of the International Sociological Association Research Committee 43—Housing and the Built Environment, Arlington, VA.

Peterman, W., & Hannon, S. (1986). Influencing change in gentrifying neighborhoods: Do community-based organizations have a role? *Urban Resources, 3*(3), 33-36, 54.

Pierce, N. R., & Steinbach, C. F. (1987). *Corrective capitalism: The rise of America's community development corporations.* New York: Ford Foundation.

Piven, F. F., & Cloward, R. (1993). *Regulating the poor: The functions of public welfare.* New York: Vintage.

Prestby, J. E., & Wandersman, A. (1985). An empirical explanation of a framework of organizational viability: Maintaining block organizations. *Journal of Applied Behavioral Science, 21,* 287-305.

Reardon, K. M. (1997). Participatory action research and real community-based planning in East St. Louis, Illinois. In P. Nyden, A. Figert, M. Shibley, & D. Burrows (Eds.), *Building community: Social science in action* (pp. 233-239). Thousand Oaks, CA: Pine Forge.

Research and Policy Committee. (1995). *Rebuilding inner-city communities: A new approach to the nation's urban crisis.* New York: Committee for Economic Development.

Riger, S., & Laurakas, P. J. (1981). Community ties: Patterns of attachment and social interaction in urban neighborhoods. *American Journal of Community Psychology, 9,* 55-66.

Ross, B. H., & Levine, M. A. (1996). *Urban politics: Power in metropolitan America.* Itasca, IL: Peacock.

Rubin, H. J. (1993). Community empowerment within an alternative economy. In D. Peck & J. Murphy (Eds.), *Open institutions: The hope for democracy* (pp. 99-121). Westport, CT: Praeger.

Rubin, H. J. (1994). There aren't going to be any bakeries here if there is no money to afford jellyrolls: The organic theory of community based development. *Social Problems, 41,* 401-424.

Rubin, H. J., & Rubin, I. S. (1992). *Community organizing and development.* Boston: Allyn & Bacon.

Sacks, D. H. (1989). Celebrating authority in Bristol, 1475-1640. In S. Zimmerman and R. F. E. Weissman (Eds.), *Urban life in the renaissance* (pp. 187-223). Newark: University of Delaware Press.

Salem, G. (1993). Participatory politics. In D. Simpson (Ed.), *Chicago's future in a time of change* (pp. 206-216). Champaign, IL: Stipes.

Sandburg, C. (1994). *Chicago poems.* New York: Dover. (Original publication 1916)

Save the Sox—on the west side. (1988, May 30). *Chicago Sun-Times.*

Schumacher, E. F. (1973). *Small is beautiful: Economics as if people mattered.* New York: Harper & Row.

Schwartz, A., Bratt, R. G., Vidal, A. C., & Keyes, L. C. (1996). Nonprofit housing organizations and institutional support: The management challenge. *Journal of Urban Affairs, 18,* 389-407.

Sennett, R. (1976). *The fall of public man.* New York: Knopf.

Smith, C. (1995). *Urban disorder and the shape of belief.* Chicago: University of Chicago Press.

Southworth, M., & Ben-Joseph, E. (1995). Street standards and the shaping of suburbia. *Journal of the American Planning Association, 61,* 65-81.

Speeter, G. (1978). *Power: A repossession manual.* Amherst: Citizen Involvement Training Project, University of Massachusetts.

Spielman, F. (1988, April 19). Sox park "can be saved": Firm revises assessment of Comiskey. *Chicago Sun-Times,* p. 8.

Squires, G. D. (1994). *Capital and communities in black and white: The intersections of race, class and uneven development.* Albany: State University of New York Press.

Squires, G. D., Bennett, L., McCourt, K., & Nyden, P. (1987). *Chicago: Race, class, and the response to urban decline.* Philadelphia, PA: Temple University Press.

Squires, G. D., & Velez, W. (1987). Neighborhood racial composition and mortgage lending: City and suburban differences. *Journal of Urban Affairs, 23,* 217-232.

Squires, G. D., Velez, W., & Taeuber, K. E. (1991). Insurance redlining, agency location, and the process of urban disinvestment. *Urban Affairs Quarterly, 26,* 567-588.

Stein, C. S. (1951). *Toward new towns for America.* Liverpool, UK: University Press of Liverpool.

Stoecker, R. (1997). The CDC model of urban redevelopment: A critique and an alternative. *Journal of Urban Affairs, 19,* 1-22.

Suttles, G. D. (1972). *The social construction of communities.* Chicago: University of Chicago Press.

Turner, M. A., Struyk, R. Y., & Yinger, J. (1991). *Housing discrimination study.* Washington, DC: Urban Institute.

United Way/Crusade of Mercy. (1996). *A best practice study of affordable housing property management.* Chicago: Author.

United Way Needs Assessment Committee for Community Development. (1998). *A new strategy for community development in Chicago.* Chicago: United Way/Crusade of Mercy.

Vander Weele, M. (1994). *Reclaiming our schools: The struggle for Chicago school reform.* Chicago: Loyola University Press.

Vidal, A. C. (1992). *Rebuilding communities: A national study of urban community development corporations.* New York: Community Development Research Center, Graduate School of Management and Urban Policy, New School for Social Research.

Vidal, A. C. (1997). Can community development re-invent itself? The challenges of strengthening neighborhoods in the 21st century. *Journal of the American Planning Association, 63,* 429-438.

Weiner, D. H. (Ed.). (1993). *The Chicago affordable housing fact book: Vision for change.* Chicago: Chicago Rehab Network.

Werth, J. T., & Bryant, D. (1979). *A guide to neighborhood planning.* Washington, DC: American Planning Association.

Wilson, W. J. (1987). *The truly disadvantaged: The inner city, the underclass, and public policy.* Chicago: University of Chicago Press.

Wiltz, T. (1990, January). Neighborhoods: Battle over new Comiskey Park goes into extra innings. *Chicago Enterprise,* pp. 14-15, 28.

Worley, W. S. (1990). *J. C. Nichols and the shaping of Kansas City: Innovation in planned residential communities.* Columbia: University of Missouri Press.

# INDEX

# ABOUT THE AUTHOR

William Peterman is Professor of Geography at Chicago State University, where he is coordinator of both the Fredrick Blum Neighborhood Assistance Center and the Calument Environmental Resource Center. Both centers work with local community organizations, agencies, and governments to focus the resources of the university for the purpose of community problem solving. Dr. Peterman received his Ph.D. from the University of Denver, specializing in the area of urban geography. He has had 25 years of experience creating and maintaining university/community collaborations. He previously was associate director of environmental studies at Bowling Green State University and was director of the Voorhees Center for Neighborhood and Community Improvement at the University of Illinois at Chicago. He has been the senior faculty fellow for the Illinois Campus Compact for Community Service and has received community service awards from the Chicago United Way/Crusade of Mercy, Chicago's Leadership Council for Metropolitan Open Communities, and HOPE Fair Housing Center of suburban Chicago.